PHOTOGRAPHY *by* N. JANE ISELEY

TEXT *by* WILLIAM P. BALDWIN

PRODUCTION COORDINATION *by* ALICE TURNER MICHALAK

Lowcountry Plantations Today

LEGACY PUBLICATIONS

A Subsidiary *of* Pace Communications, Inc., Greensboro, North Carolina

www.pacecommunications.com

We wish to dedicate this book to William P. Baldwin, Jr., and his wife, Agnes Baldwin, who taught us to appreciate the beauty and value of the Lowcountry and its plantations.

PHOTOGRAPHY © 2002 / N. Jane Iseley   TEXT © 2002 / William P. Baldwin

ACKNOWLEDGMENTS

This book would not have been possible without the kindness and generosity of the many owners of properties, both private and public, who graciously allowed us to visit and photograph. We are deeply grateful for their cooperation and hospitality. – N. Jane Iseley, Alice Turner Michalak

———————

In 1984, my mother, Agnes Baldwin, and I did the text for Jane Iseley's *Plantations of the Low Country*. Once again, it was a pleasure to work with Jane, and though this time my mother's name is not on the title page, I am still in her debt. As I wrote then, "To protect the privacy of the current owners of the homes included here, their names have been omitted from the text. This anonymity is certainly no reflection on their contribution." Again, the owners and their managers and staff have been both gracious and helpful. / Designer Jaimey Easler, editor Sheryl Miller, and that ultimate "push" Boots Michalak can't be thanked enough. With great skill, they manage to smooth away a multitude of sins. / As always, the Charleston Library Society, the Charleston County Public Library, the Village Museum, and the South Carolina Historical Society proved their worth. Most particularly, I'd like to thank the following librarians and archivists: Bud Hill, Catherine Sadler, Pat Bennett, Dee Syracuse, Janice Knight, LeeAnn Floss, Alex Moore, Harlan Greene, Peter Wilkerson, and Pat Gross. / Books . . . I am indebted to the following classics: Sam Stoney's *The Carolina Low Country*, Alberta Lachicotte's *Georgetown Rice Plantations*, and Todd and Hutson's *Prince William's Parish and Plantations*. And I referred often to and cited from two recent and indispensable contributions: Virginia Beach's *Medway* (Wyrick and Co.) and Suzanne Linder's *Historical Atlas of the Rice Plantations of the ACE Basin 1860* (South Carolina Department of Archives and History). / Conversations . . . Over the last three years, I enjoyed listening to dozens speak on the beauty and usefulness of today's plantations. Thanks to all of you, especially furniture restorer Robert Sareo, architect Virginia Lane, and most especially, landscape architect Robert Marvin. A remarkable man, Robert died while this book was going to press, and I expect the interview included here is one of the last he gave. Since he was ill even then, I intended to stay at his home for only a few minutes. We spoke for two hours. And parted. Then as I was backing the car from his drive, I heard my name called out. Robert was standing in the doorway, a bent, frail man, but one still capable of a broad smile. "Remember, Billy," he said. "The land comes first." – William P. Baldwin

DESIGN / Palmetto Graphic Design Company   EDITOR / Sheryl Krieger Miller   PROOFREADER / Sarah Lindsay

ISBN: 0-933101-21-X   LIBRARY OF CONGRESS CONTROL NUMBER: 2001097418

PREPRESS / NEC, Nashville, Tennessee, USA   PRINTING / D.W. Friesens, Altona, Manitoba, Canada

LEGACY PUBLICATIONS / 1301 Carolina Street, Greensboro, North Carolina 27401

The Hounds of Twickenham

Compared to centuries past, today's Lowcountry plantations produce very little—very little beside beauty and enjoyment. Yes, times have changed. Along this 200 miles of South Carolina coast, these venerable, but once neglected, country places have been refurbished and renewed. Usually approached down avenues of ancient live oaks, the handsome, even stately, houses are set on gentle rises.

Their great expanses of green lawn are banked with azaleas that glow in springtime colors of pink, purple, and crimson.

Often measuring in the thousands of acres, the surrounding land is well-tended. The pines, oaks, and hickories have reached impressive heights, and the forest floor is manicured by burning. The cypress trees grow thick in the black-water recesses. Rice fields, home now to a host of migratory waterfowl, are drained, planted, and flooded. And other wildlife habitats are encouraged and protected.

The natural world is unchallenged. Today's plantations are places of quiet occupation, orderly pleasures, and sheltered wilds. And the plantation houses, many of them of antebellum origin, receive the same reverent attention. Paint has replaced whitewash or no paint at all. Roofs have been repaired, and broken windows replaced. Electricity and plumbing have been installed. Sunrooms added. And previously empty rooms are now furnished with elegant pieces. And most recently, an entirely new "Lowcountry architecture" has brought brightness and harmony of a more modern nature—without abandoning the more traditional, more familiar, aspects of plantation building.

But most important, whether old or new, the plantation homes of today are free from the constant threat of deadly disease and the tensions detrimental to domestic security. Now, at last, these houses are truly commodious and firm. They are livable.

In 1932, historian Sam Stoney remarked, "Life in the Low Country has never lacked a certain dramatic intensity of mood." His reference was in part to the bold nature of South Carolina's coastal landscape, but also to the exotic existence of its occupants—the romance of a planter caste touched by "the imperial color of purple," then following the Civil War, reduced and dispersed from their land. And yet, viewed in the light of modern thought and scholarship, from the very beginning these plantations were linked to tragedy—to death, enslavement, and defeat. In short, to the romance of unavoidable violence and loss. And this ongoing drama was hardly applicable to a comfortable or even fruitful existence.

Viewed in that light, how much happier is the present situation.

HERE AND ELSEWHERE

When settlement began, plantations were everywhere. Any parcel upon which seeds were planted qualified as a plantation. The man who worked this land was a planter. When reluctant Scots were relocated in Northern Ireland, the place was called Londonderry Plantation. When the Pilgrims landed in the New World, they established Plymouth Plantation, and each settler held his own smaller plantation. But gradually the term came to be associated only with relatively large agricultural enterprises that produced one crop and relied on ganged labor. At the earliest, this was often slave labor. These working plantations are still around. In the tropical and subtropical latitudes they circle the world producing coffee, rubber, bananas and more. But for most Americans today, the word is synonymous with our own Southern plantations, those romantic, if dark, vestiges of pre-Civil War America, those lands that now serve as a buffer against this country's—against the South's—rising flood of industrialization and settlement.

We, however, are concerned only with the Lowcountry, a 50-mile-wide strip of coastal plane running from Georgetown to Savannah. It's a narrow stretch of Southern land and is atypical, for here the plantation culture was well-established by 1700; much of the South would only prosper with the introduction of the cotton gin a century later.

Along the Ashley and Cooper Rivers | *Harper's New Monthly Magazine, 1875*

At the confluence of the Ashley and Cooper rivers, Charleston was founded. From there, the land was taken up in every direction. The first settlers depended on the Indians for food and protection, and almost immediately they looked to them for commercial gain as well, trading for deer skins and other Indians. And soon hogs and cattle were let loose to roam freely and the forest was cut and tapped for ship stores. But already surveyors were busy setting out property boundaries and a great variety of agricultural experiments began.

For the newcomers, land meant not only wealth; its possession was the key to social order, a principle actually written into the colony's charter. The eight Lord Proprietors intended to divide Carolina into vast 12,000-acre baronies that would be sold or given to noble landgraves and cassiques, the colonial equivalents of earls and lords. The less fortunate would receive far less property. The plan failed, for though birth and breeding still mattered, the rigors of settlement seldom suited the lifestyle of a true country gentleman.

From England and the West Indies some younger sons of the aristocracy arrived with their notions of feudal feifdoms. But Puritans from both England and New England came as well, dissenters who brought with them a notion of Old Testament pastoralism—with a democratic twist. Scots and Irishmen and fiery Scot Irish came, too, being pushed from their homes by the English and angrily expecting a new chance. And French Huguenots, driven from their middle-class urban homes by religious persecution, settled in the midst of wilderness. They all came. Disease quickly carried away many, and from those who remained, the most daring and quick witted had within a short time created a new aristocracy of survivors. Property was still the basis of wealth and power. The raw land had become "plantations" and the settlers were now true "planters."

As mentioned, the concept of plantation was hardly unique to the Lowcountry. On Barbadian estates, slaves were growing sugarcane and in Virginia, tobacco. All Carolina needed was a crop. Olives, grapes, and silk were among the experimental crops, but rice proved to be the answer. Beginning in the early 1700s, inland swamps were cleared and diked— reservoirs provided the water for cultivation. Then, after the American

Revolution, the marshes and swamps directly bordering the rivers came under cultivation—the ocean tide pushed fresh water into well-ditched fields through special water-control devices called trunks. Following the Civil War, this elaborate hydraulic system would lend itself equally well to the maintenance of "duck marshes."

In about 1745, indigo was introduced as an upland crop. The most notable of the first experimenters was a 16-year-old Barbadian girl, Eliza Lucas. Indigo produced a dye so valuable that the English offered a bounty to keep it out of French hands. Fortunes were made—until the colonists fought the English in the Revolution. Soon after, the invention of the cotton gin brought prosperity to the more inland properties, and the introduction of "Sea Island" cotton did the same for those along the coast. As with rice, crops cotton was labor intensive, and without forced labor, the "grand" profit was gone. Cultivation did continue, however, until the boll weevils' assault. These fields were then planted in vegetables or allowed to return to forest. When walking up a covey, today's quail hunters are sometimes treading on these almost forgotten rows.

When speaking of plantations, this post-bellum (after the Civil War) career is often neglected. This is unfortunate, for in many ways, today's plantations are more productive or less destructive than in centuries past. Hunting had always been the joy of the early planters. Deer, most especially, were pursued from horseback and driven to hunters waiting on "stand." But quail, fox, snipe, turkey and much else were also sport. "Whatever crossed our rambles," as one young nimrod put it. Oddly, though, ducks and geese were seldom thought more than a larder source—perhaps, because they were so numerous.

After the War, with the plantations in ruin and bankrupt, this was to change. By 1890, Yankee sportsmen were well-established in the lower corner of the state—at a time when it was still dangerous to be a Yankee. The abandoned rice fields were proving to be excellent duck marshes, and an overnight train ride from Philadelphia or New York brought the hunter to a hunter's paradise. Gun clubs, groups of like-minded investors, were quickly formed over the next decade. Plantations were bought up and redirected. Soon after, private individuals were doing the same—a process that slowed after the turn of the century and revived again with the coming of the Great Depression.

Writing for *Country Life Magazine* in 1932, James Derieux described the attraction of the Lowcountry plantations to their new owners: "It was the hunter who really 'discovered' the charms of the Carolina coast. He found there plenty of deer, turkeys, ducks and quail. He invited his friends, and they in turn bought places for themselves. They invited their friends, and thus the redemption of the Lowcountry got underway." And spurring on this migration from the other direction were the pressures of the "push and shove" Northern life. "There is no place where the sound of a stock ticker would seem so ill suited to the [plantation's] surroundings."

As Virginia Beach, historian and environmentalist, has pointed out, the median per capita income in South Carolina in 1929 was a dismal $269, illiteracy was the highest in the nation, and unemployment figures were in grim accordance with the rest. Still, from the viewpoint of the native-born Southerner, the hunters' intrusion has not always been viewed happily. But it cannot be denied that the quail, duck, and deer hunters did provide a new service-oriented economy to be formed around the plantations, and of equal importance, tremendous acreages were preserved in relatively pristine condition.

So today, while the rest of the world seems to be hurtling toward endless strip malls and repetitive suburban communities, nature preserves of astonishing numbers and size are being established in the Carolina Lowcountry. Plantations are gradually passing into the hands of state and federal refuges and like-minded conservation trusts, or private owners are using easements, either alone or collectively, to create similar havens for both beasts and man. After all, as one landowner remarked, "It's the considerate hunter who is the true conservationist and the owner of property who most often values the natural world of which he can never be more than a custodian." Admittedly, the amount of game is sadly decreased, but the pleasure of simply being in the wild remains constant, and if we are to look for a true Golden Age of Plantations, we should not be looking to the past but to this very moment.

A LOOK BACKWARD

However, it is natural to look backward for that romantic, if elusive, time of better things. Until the coming of the Civil War, many planters prospered. The most fortunate of them enjoyed the finest wines, thoroughbred horses, and handsome furniture in handsome homes. The sons had the best education and the daughters the prospect of good marriages. Coachmen drove liveried carriages. The libraries were grand, the conversation sparkling. In these plantation homes, an 1804 visitor wrote, "You meet the polish of society, and every charm of social life; an abundance of food, convenience and luxury." And undeniably, the leadership in the founding of this nation often came from the Southern planter class. (Some historians manage to deny it.) George Washington and Thomas Jefferson come quickly to mind, but the Carolina Lowcountry provided dozens of soldiers and statesman who did almost as much.

WORLDS UNTO THEMSELVES

Plantations, at least the larger ones, were little worlds unto themselves—a mansion for the owner and his family, often some flanker buildings, and always a kitchen at some distance from the house. Close by were a smoke house and creamery or icehouse or spring house, and somewhere a privy. The overseer had a house of his own, and the slaves were settled on a "street" of small single- or double-room cabins. In addition, there were sometimes an infirmary and praise house for the slaves, workshops for carpenters and black-smiths, and of course, stables and pasture for livestock, a rice mill or cotton gin, a winnowing house, great storage barns, and a river dock or landing. Beyond that were the fields. Today, no plantation exists completely intact, but the layouts of Boone Hall and Middleton Place suggest what once was, and most every plantation has earlier outbuildings converted to modern uses.

Often as much care went into the grounds surrounding a plantation mansion as into the house itself. From the earliest days, local scientists excelled in the study of botany—and the pursuit of gardening. It helped to have a solid foundation, to have the rich, shiny green gift of a Florida maritime forest or its inland version to build on. Live oaks, palmetto, and more exotic greenery were on site. And sea captains circling the world could be counted on to bring treasures from the world's Far East.

Fashionable in England, the tree-lined avenue was easy to come by in Carolina. Live oaks lent themselves readily, but cedars and magnolias were also early favorites for entrances. The gardens themselves were also begun along Old World lines—formal, geometric, and trimmed—but by the 1840s, camellias and azaleas were becoming common, and a more romantic approach to plants and their arrangement began. (Many of the oak avenues also date from this period.) At Magnolia Gardens and Middleton Place, we see an excellent example of each garden type done to a grand scale. On many other plantations, similar gardening still goes on. In the spring of 1875, one Lowcountry visitor exclaimed: "Stand on one side and look across the lawn; it is like a mad artist's dream of hues; it is like the Arabian nights." And then she added: "Sometimes in the Northern gardens one may see, carefully tended, a little bed of scarlet geraniums all in bloom, or else a mound of verbenas in various shades; imagine these twelve or thirteen feet high and extending in long vistas in all directions as far as the eye can see." But in fairness to our Northern brethren, it should be said that plantation gardens were also kept (and still are) on the more personal scale alluded to. As Charleston's master gardener Emily Whaley explained, a garden (or a portion of the garden) should be enclosed, should "push back the wilderness ... be an intimate ground safe from lions and elephants and whatever else is out there."

Still, we must bear in mind that a garden of any size is far more ephemeral than a house—part of its appeal perhaps. Of the more than 300 gardens that Emily's own noted designer, Lotrel Briggs, did in the Lowcountry just a half century ago, only three dozen or so remain, and these are in altered form. Our landscape's fecundity leads to endless alteration, and unless a firm brick pattern has been laid, even the shape must be surrendered to the changes of nature and fashion.

A PECULIAR INSTITUTION

From its beginning, Carolina, and the Lowcountry in particular, was not viewed with kind eyes—most especially when slavery was the issue. Bishop

Asbury, "the father of American Methodism" traveled here often after the Revolution and wrote, "How much worse are the rice plantations! If a man-of-war is a 'floating hell,' these are standing ones." Similar accusations were made by early Quakers and Baptists and then loudly voiced by Northern abolitionists.

If the plantations did succeed in becoming romantic little worlds unto themselves, theirs was not a romance of ease shared by all. They were populated overwhelmingly by slaves—slaves who, to many master's amazement, were not content with their station in life. The Reconstruction period following the Civil War was one of chaos. Blacks eventually moved away or organized tight self-reliant communities of their own. But some remained on the plantations and continued to enrich these places with their unique African-American-Creole culture.

AND THE COUNTRY FEVERS

Slavery wasn't the only dark cloud over the plantation Lowcountry. Malaria, then called "country fever," was another. "Miasma," or night vapor, thought to be the first cause for the illness, was quickly associated with swamps and rotting vegetation. Soon plantation houses were built high above the ground and as far from the swamps as practical. (Today we know that the female Anopheles mosquito transmits the disease, a mosquito that most often flies close to the ground and at dusk.) Still the fevers persisted, and whereas a healthy adult might survive and even build up an immunity, it was easily fatal for the very young and the old. For this reason, planters began to retreat from the land during the warmest six months of the year, staying in small seaside or pineland villages or even in distant resorts. And it should be noted, many of today's owners still flee at times to more comfortable climes.

ARCHITECTURE

In the Lowcountry, about 150 antebellum plantation houses remain. Most can be visited at least once a year on tours sponsored by churches or historical societies. With a strong tradition of craftsmanship, they're usually well built. Local long-leaf pine went into the heavy beams, and weather-proof black

Duck Hunting on Lowcountry Rice Fields | Harper's New Monthly Magazine, 1878

cypress shielded both the inside and outside. Often made on the site, brick was the early builder's choice for walls, but gradually wood siding came to be preferred. Foundations were always of brick—except when an oyster-shell-based concrete called tabby became popular. The roofs were of cypress shakes, tiles, or slate. On the verge of the Civil War, soldered tin appeared. All of the above are still in use—even on the newest of houses. Judging by the records, white was the most frequent exterior color and still is today.

The majority of the Lowcountry's remaining antebellum plantation houses and the majority of those shown in these pages may be loosely classified as

"Georgian." With the exception of Drayton Hall, most are relatively modest examples. This 18th-century design stressed symmetry: A typical English dwelling was two story with a low-pitch roof and a small central portico entry, but the demands of the Lowcountry climate required higher ceilings and often verandas across the front. The interiors were paneled, the moldings thick, and the windows small and small paned. The floor plan that was most often used was the traditional central core of four rooms over four with a central hallway dividing each floor, an arrangement popular even now.

Next came the Federalist period, that 30 years following the Revolution when the influence of England's Adam brothers was felt. Both inside and out, moldings were more delicate and curved Palladian windows, trim double porticos, and protruding bays suggested a new refinement. Finally came the Greek Revival, those great columned mansions that coincided with the boom that cotton planting brought to the rest of the South. Though this is the image most often associated with Southern plantations, actually few were built in the Lowcountry.

Styles change slowly here, and Georgian was never completely abandoned. In the 1890s, when architects began to design a new series of plantation homes, it was most often the Georgian or Federalist country place that they turned to for inspiration and not the Greek Revival of the South's more western regions. This is especially true in the Beaufort area, where Sherman's troops burned practically every house in their path and left behind beautiful live oak crowded sites and nothing else. And since the new wave of more constructive Yankees followed in his wake, it's not too surprising that they sometimes built hunting lodges in the New England "shingle" or even the Adirondack tradition. It was a rich mixture; even Frank Lloyd Wright made a wonderful "organic" contribution. Indeed, architects designing our most recent houses seem to be looking to that master for the inclusion of light and organization of space while taking their details from a variety of earlier styles.

PRESERVATION

The national preservation movement began in the South, and Charleston's efforts to protect its historic fiber have been a model for many other cities. Quite naturally, the entire Lowcountry has been affected. As you'll see here, though empty of furniture, both Hampton Plantation and Drayton Hall are maintained in pristine condition, and on a private level, similar work is being done. As a modern convenience such as plumbing is added, the installation is done in an unobtrusive manner. When an earlier wall treatment is too shabby to be seen, it's preserved beneath a protective film before newer paints or varnishes go on. As with the maintenance of the land, such efforts are endless. But then, so are the rewards.

FURNITURE

"Carolina appears to be English," wrote an early commentator. While the planters, no matter what their origin, might look to that mother country for their style of furnishings, we should remember that the very concept of furniture was in one sense new to the English. Before the Restoration of Charles II in 1670, fine furniture had been enjoyed for the most part only by the royalty. The same "democratic and capitalist spirit" that was needed to organize the settling of Carolina also helped to introduce a conglomerate of French-, Dutch-, and Chinese-designed furniture into English homes of a rising merchant and professional class. Indeed, it's thought that the French Huguenot settlers who moved into England in the 17th century had a major impact on the English furnishings that our own French Huguenot planters purchased only a couple of generations later.

Chippendale, Hepplewhite, Sheraton. The 18th century has been called the "Golden Age" of English furniture. And the same term has been applied to the entire Carolina Lowcountry at that time. Although the English craftsman were best known because they published their designs, museum expert Milby Burton pointed out that Charleston's Thomas Elfe not only built furniture every bit as good as his English contemporary Chippendale, he charged more for it. In 1810, 81 cabinet makers were practicing in Charleston and at least a handful more were working in Beaufort, Georgetown, and Savannah. These craftsmen, with their frequent advertisements that "orders from the country will be punctually complied with," were the source of plantation furnishings—at least the most finished pieces. More primitive "Plantation made" cabinets and tables, usually of cypress or pine, were being made on site by slave carpenters—which is not so much a reflection on their abilities, since famed Thomas Elfe and most of the other Charleston cabinet makers were using slave artisans as well.

In his catalog, "The Gentleman and Cabinet Maker's Director," Thomas Chippendale actually worked in three basic styles—Gothic, rococo, and Chinese—a broad range not always appreciated by collectors. Usually he built with mahogany—the designs were lighter and the carvings more intricate but less lavish. Of course, it's chairs that he's most often associated, chairs whose graceful cabriole legs ended with claw and ball feet. But his tripod-based tables were almost as popular. Hepplewhite, whose guides were published following the American Revolution, brought an even greater simplicity and delicacy to design—notions well suited to our Adam brothers-influenced Federal period. He's noted for his graceful (and comfortable) shield and oval backed chairs and for elegant sideboards and other bow- and serpentine-fronted furnishings, much of it inlaid or painted. And in the Lowcountry, his suggestion that furniture be taller or "long waisted" to accommodate the higher ceilings was particularly helpful. Bedposts of 9 feet were fairly common. And finally came Sheraton, who produced furnishings that were even more fragile in appearance and relied even more on inlaid patterns and painting.

Chippendale, Hepplewhite, and Sheraton were "translated" for the Lowcountry by craftsmen who either came from or were trained in numerous European cities. They served a clientele that was equally cosmopolitan. Naturally, American masters like Charleston's Elfe made substantial contributions of their own, but the three English craftsmen are mentioned because in the following pages we'll often see if not their actual furniture, their influence. Also, as Burton points out, few pieces of Charleston-made furniture actually survive. Numerous invasions, fires, poverty, and wear and tear have taken their toll. Still, as knowledgeable furniture restorer Robert Sarvo can attest, Charleston pieces are being preserved with considerable care and used with pride. "Charleston pieces," he says, "are the fun ones." But with state-of-the-art craftsmanship and wonderful proportions, all work of the period is exceptional, and many of today's plantation residents have been drawn to that 18th- and early 19th-century elegance—an elegance that is a natural complement to houses whose design and construction grew out of the same cultural milieu. And, of course, many residents are happy to mix in reproductions from the 19th and 20th centuries—and a modern piece or two. Indeed, Frank Lloyd Wright-created furniture is found in the houses he designed for Auldbrass, and no doubt, more contemporary house designs will lead eventually to more contemporary furnishings. Still, much can be said for those 18th-century pieces—for all that functioning grace and all that rich-red mahogany.

THEN AND NOW

There are still a few ancient souls around who remember the South before air-conditioning and massive applications of bug spray. That was a different South, one that the planters and their families sometimes abandoned completely during the warmest months. Nevertheless, the houses they built were designed to stay cool. The ceilings were high and the windows generous. The planters could put up with the cold. Fireplaces were centrally located, firewood plentiful and clothing thick and warm. So if we are to look for the most obvious changes in living, they would be in the mechanical accomplishment of a year-round comfort zone. Central heating and air-conditioning are in place. Of course, the high ceilings and tall windows, now screened, still can be counted on for cooling and they provide a handsome design dividend as well. And those high foundations, with the porches oriented toward the prevailing breezes and the river, create comfortable spaces with commanding vistas.

Lighting, too, has created a broader time frame for occupancy. Imagine, if you can, only candles burning—the flames of these and perhaps an oil lamp or two reflecting against the mirrors, the shimmer and romantic mystery, and the absolute inconvenience of finding your way around in the semi-darkness. But at the same time imagine a world without telephones—without faxes and deadlines dictated only by the methodical passing of the seasons, the life inside a house structured only by the rising and setting of the sun. (But then, Burton lists 50 clockmakers working in Charleston in the 18th century.) Imagine entertainment based on the family's own ability to perform, and dramas delivered through books instead of satellite-born talking images. But these same "modern" pressures have brought a new understanding of just how precious our connection to nature is. As landscape architect Robert Marvin points out, even with their generous windows, most Georgian and Greek Revival architecture was no more than a box for locking nature out, an adversarial relationship not only with flora and wild beasts but with light itself. It's not uncommon these days for curtains to be of the thinnest fabric possible or even nonexistent and for the shrubbery and trees closest to the house to be trimmed away allowing a maximum of daylight to enter. The most recently built plantation houses all have walls of glass, and several of the older houses have been opened up on their less formal sides with great windows and sun or garden room extensions.

Perhaps the other most significant change would be in the number of inhabitants. Even with an alarming high mortality rate, families were large and were made larger still by the extended inclusion of numerous aunts, uncles, retired priests, orphaned children, tutors, and endless overnight guests. To this we must add from six to perhaps two dozen servants, most often slaves. Cooks, maids, nurses, a butler, even footmen were on hand. True, narrow hidden servant stairs sometimes connected the basement to the attic, suggesting, at least, an out-of-sight maintenance. And true, the kitchen was most often separated from the house because kitchens had a habit of burning down. Cooking was done on large open fires with spits for roasting and baking in the adjoining brick ovens.

These brick-floored outbuildings and their nearby spring houses and smoke houses contrast with the 21st-century conveniences now enjoyed. Equally dramatic are the changes that plantation kitchens have gone through. With two or three servants taking the place of dozens (and sometimes no servants at all), the kitchens, now moved to the main house, have often become "family" rooms and domestic offices. In the most modern of interiors, they are even connected openly to the more formal areas of the house.

In most houses, even those of post-bellum construction, these same formal areas have changed the least. Broad entries containing elaborate stairways were the norm, rooms that created a welcoming, if sometimes grand, first impression. These entryways oriented the visitor first to the immediately adjoining parlors and dining rooms but suggested a degree of intimacy to be had in the depth of the house. They still do. Even today, these most immediate areas of the houses remain the most elegantly decorated. Hospitality is the goal. Large double doors slide open to create spacious areas for entertaining, and lengthy dining-room tables accommodate dozens, just as they did when constructed two centuries earlier.

On a more relaxed level, though, it's the gun rooms, garden rooms, sunrooms, libraries, dens and family rooms where true unwinding occurs. These are placed a bit off the beaten path. At the end of halls or through adjoining rooms, they are comfortable but still well-appointed spots where coffee may be drunk and the ears of dogs scratched.

Bedrooms. In early days, the bedstead was a most valuable furnishing. In fact, the most common listing in the early Lowcountry wills is of chairs and bedsteads. Plantation bedrooms today are handsomely turned out. True, a canopy of gauze mosquito netting hanging from the four posters is no longer a life and death matter. True, the bed is no longer pulled midroom in summer and the head removed to increase circulation. And warm bricks and bed warmers have passed from fashion. But today's bedrooms are well-appointed retreats from life's waking demands. Fireplaces and easy chairs add another dimension as does the spring morning view from the second-story window. And the guest bedrooms are often as carefully created as the master suite.

Closets. There once were none. In many houses today, cabinets of various sorts still serve as replacements. Bathrooms. There once were none. Chamberpots and nightstands took their place, and outside were privies. Bathing was far less frequent and done in temporarily provided metal tubs.

The actual placement of furnishings has changed, as well. Before 1760, furniture was placed along the walls, leaving the center of the room free. As the docents explain at Drayton Hall, the room was "straightened." But by the end of the Revolution, the new, lighter, more delicate furnishings were grouped in the center of the room, with tables, chairs, and settees organized with a far more casual comfort in mind. Gradually, this trend in both delicate structure and comfortable arrangement was extended. Combined with lighter wall colorings and opened curtains, it gave the rooms a more relaxing atmosphere. Curiously, this sense of "home" had by the mid-1800s given way to the dark walls, heavily turned furniture, and endless bric-a-brac that we associate with the Victorian era. And though we don't usually associate this trend with plantation houses, on the eve of the Civil War, many were decorated in just this manner. Modern times have brought a pleasing return to more gracious and livable interiors.

But as already suggested, if we are to judge plantation homes by both their beauty and their comfort, we should begin to consider that the plantation's "Golden Age" was not yesterday but today.

Along the Savannah River | Harper's New Monthly Magazine, 1853

AN ORDERING

The Lowcountry is cut by a meandering of creeks and rivers. This landscape didn't lend itself to building and maintaining of roads—nor did the independent nature of the planters. From the beginning of settlement and into the 20th century, the waterways served instead. The earliest houses often had their formal entries to the river side. Naturally, then, rivers became the backbones of neighborhoods. Relatives and friends settled along their length. They built houses and tended to the grounds. This strong "localism" can be found up to the present day—especially in those neighborhoods south of Charleston.

The towns were the other binding factor—for in them isolated neighbors came together for commercial and social engagements. Charleston, Beaufort, and Georgetown—all three served as ports and markets.

And so we have organized this book around both towns and rivers—beginning along Charleston's Ashley and Cooper rivers, where the earliest of houses were built, going next to the places of the Georgetown river systems, and ending with the Beaufort area, where because of the Union invasion, we tend to find those houses that are newest.

We hope you enjoy what follows. Compiling it has been a pleasure.

The city of Charleston's relationship with all of the Lowcountry has been so close, so symbiotic, that she and her surroundings have been compared to a Greek city-state. Settlement began here, and the joining of the Ashley and Cooper rivers gave her a fine harbor. All government was centered here, and the latest of European fashions—translated by furniture makers and silversmiths—came first to this wealthiest of North American cities. An energetic plantation-based society settled quickly along the relatively short courses of the Ashley and Cooper rivers.

Charleston and Her Rivers

MIDDLEBURG PLANTATION / House circa 1697, Cooper River, East Branch

Following the execution of his Huguenot parents, 15-year-old Benjamin Simons escaped France and went to the Dutch city of Middleburg. He then made his way to London and joined with his aunt and uncle to emigrate to Carolina. Once here he married Mary Ester DuPre. After gaining a warrant to 100 acres, he built this house, which was finished in 1697. Soon after, he received a grant for another 350 acres from the Lord Proprietors and began to plant rice and indigo as well as manufacturing naval stores and raising cattle.

On his death, Benjamin left the property to his youngest son, Benjamin Jr., who at age 4 was the youngest of 14 children. By then, Middleburg consisted of a considerable 1,500 acres. However, Benjamin, Jr., was later trained as a carpenter and would spend most of his life in Charleston as a businessman and not a planter. His son, Benjamin Simons, III, also worked as a craftsman and merchant. He died in 1789, leaving more than 3,000 acres to be divided between three daughters. It was at this point that the tidally influenced river became a source of rice field irrigation. Miles of new dike were added, and rice became the single crop.

Where necessary, hand-carved mullions and sills replaced the decayed ones. Thin Federalist mullions are used here. The salvaged glass is naturally mottled and cast on the avenue beyond a slightly blue tone (below). *Middleburg is the oldest house in South Carolina* (right).

In 1799, Sarah Simons married Jonathan Lucas, Jr., the son of a famous rice mill inventor and an inventor and entrepreneur in his own right. He soon built a huge rice mill at Middleburg, then several more in Charleston. When Parliament placed a tariff on rice milled outside of English territories, he returned with his wife to England and began a second illustrious business career. By this time he had added to Middleburg a Flemish-style stable (destroyed by Hurricane Hugo), a guitar-shaped garden, an ornamental pond, and the avenue of 32 oaks that we see today.

Middleburg next passed to Jonathan Lucas, III, who contracted pneumonia and died. His son Thomas inherited the plantation but lacked the experience to manage the family fortune,

An Adamesque mantel with delicate carving and an Italian marble surround.
The chair and vase seem to comment on the graceful garland et al. (left).
A nutting stone or Indian mortar dug up in the yard. It was quite usual
for Indian settlement sites to be taken up by the planters, for high ground by a river and
a convenient spring were necessary to both. The iron cooking utensils represent
the next generation in food preparation (above).

most of which was soon lost. The plantation stayed in the family's name throughout the Civil War, however, and was then sold to John Coming Ball, who was one of the few successful planters in the post-war period. He added three large tracts and married the great-great-granddaughter of Benjamin Simons, III, thus returning Middleburg to its first family.

Two great hurricanes in the 1890s put an end to the commercial growing of rice in that neighborhood. John Coming Ball, however, continued to grow and mill a small amount until his death in 1923. Then Middleburg was largely abandoned until 1941, when it became the home of Ball's daughter and her husband, the wildlife painter Edward Von S. Dingle.

The present owners made their purchase in 1981 and began a careful and extensive renovation of both house and grounds. Needless to say, from the history just recited, such a restoration was a happily conducted architectural "Who done it?" or more exactly, "What did they do and when?"

This corner cupboard was a late addition to the house. The brightest of those plates were in the house in the 1820s. Careful analysis of the paint revealed a curdled milk-white for the plaster and an oxblood-based red for the trim. Modern equivalents are on the walls today (below).

A sunny hallway on the second floor offers a glimpse of the period's furnishings. Around the corner by the stairway is the bathroom. This necessary room was built self-contained and may be totally removed if the house is ever converted into a museum (right).

As is often the case, the earliest house was small, only two rooms over two—these now located to the right. With the birth of Benjamin Simons' third child, the builders began an additions project that occupied the family for more than a century. By then the front of the house had been turned to the road, and chimneys and windows, doors and staircases all migrated with some freedom.

Though Middleburg is the oldest house in South Carolina, the owners chose to return it to a more manageable pre-1800 condition. Working in close collaboration with contractor David Hoffman, they have carried out a museum-quality restoration.

A conservation easement protects Middleburg Plantation's current 300 acres from any development.

The house is on the National Register and is a National Historic Landmark. ✥

MEDWAY PLANTATION / House circa 1705, Back River, Medway River

In 1688, the wealthy Dutch widow Sabrina van Aerssen married widower Landgrave Thomas Smith. She died soon after, leaving Smith with her first husband's claim to a barony, a plantation and brick house on the Medway River. Not long after, Smith, who was then governor of the colony, died at the age of only 46 and was buried on the plantation beside his "Dutch widow" Sabrina.

From the Smith family and completely on credit, Edward Hyrne bought "his brave plantation,"

Axis of a double avenue (below).
Two lions guard the front steps (bottom).
Two large additions have swallowed up the original humble dwelling, but the result is a harmonious whole (right).

2,250 acres of land, 200 of this cleared and fenced, plus "150 head of cattle, 4 Horses, an Indian Slave almost a Man, a few Hogs, some Household stuff, and the best Brick-House in all the Cuntry; built about 9 years ago, and cost 700 pounds, 80 Foot long, 26 broad, cellar'd throughout."

But Edward and his wife Elizabeth were hardly prepared for the rigors that faced them. Edward, a Norfolk merchant, was fleeing England ahead of his creditors and additional accusations that he had misapplied 1,300 pounds of the crown's money. Elizabeth was the 18-year-old daughter of a baronet waiting on an inheritance. Tragedies piled upon tragedies. "On the 20 of the same instant," she writes, "we lost a Negro Man by the bite of a rattlesnake which was a great lose to us being just in the height of weeding rice. On the 25 of August I lost my Dear little son which went very near to me. In September we lost our Cattle hunter. But the greatest of losses (except my dear Harry) was on the 12 day on January last on we was burn out of our house taking fire."

From the flames they had barely managed to save their remaining son and "two beds just to

previous spread / *The ancient home of the unlucky Hyrnes* (page 24, top).
*This geranium-studded patio doubles as an entry stoop. In the distance, the distinctive double
avenue of oaks planted by the Stoney family before the Civil War* (page 24, bottom).
*Greyhounds sport. "Living was rapid but there were bonuses for the astute," wrote Mrs. Stoney of
Landgrave Smith, who lies buried on this same rise* (page 25, top). *The serpentine wall was
inspired by a Thomas Jefferson creation* (page 25, bottom).
this spread / *By the 1930s, Grandmother's Garden was lost to the wilderness.
Gertrude Legendre recalled that a landscape gardener who was a friend of her mother's
"came down to Medway and on her hands and knees redefined the walks"* (above).
A *view of the kitchen garden from the potting room* (right).

lye on." All clothes and provisions were lost. Still, they persevered, building a new and smaller house on the site. But when Edward returned to England, he was thrown in prison. And in 1711, Thomas Smith II took possession once more. Smith then married a daughter of Hyrne's by a first marriage. "Living was rapid," wrote Mrs. Stoney, "but there were bonuses for the astute."

Medway's acreage would over the centuries swell and shrink and swell again. And the Hyrnes' small brick house would form the nucleus of a grand mansion. Rice-growing and brick-making were the source of profits—sometimes limited.

Dutch architect Ides Van Der Gracht designed the enclosed "kitchen garden" in the '30s (below). This old kitchen building now serves as a guest house (bottom). This Chippendale style "work" or "plantation" desk is being reproduced by the Historic Charleston Foundation (right).

In 1833, Gaillard Stoney purchased the place. It was then that the largest of the wings were added, the grounds expanded, and the double oak avenue planted. As elsewhere, the losses of the War were never recouped, and in 1906, Gaillard's nephew, Samuel Gaillard Stoney, bought the property. His son, Sam Stoney— historian, writer, architect, and storyteller— would eventually become the premiere celebrator of the Lowcountry's plantations.

But Medway's next and final chapter begins in 1929 when a young couple, Sidney and Gertrude Legendre, came visiting at neighboring Cypress Gardens and went out for a horseback ride.

"As we neared Medway, I remember peering under a festoon of gray moss dripping from huge oak trees and catching sight of an old, eerie pink structure," recalled Gertrude Legende. "Inside, mattresses covered the floor, and deer horns adorned the walls. There was no water, no light, no heat."

The young couple were certain that city life was not for them. This was the place. And against the advice of others, they bought the property from the Stoney family. With the aid of Charleston architect Albert Simons, they

This spacious dining room was an 1855 addition. Ladder-back Chippendale-style chairs

surround the banquet table, and on the far wall a Hepplewhite sideboard. The portrait is by the

English artist Simon Eloise (left). **F**or a 1930's renovation, this handsome cypress paneling was

borrowed from a collapsing plantation house next door. Library steps second as an end table.

As to the rest: "I like a place to be comfortable," explained Gertrude Legendre,

"I'm not the formal type. I don't want to worry where the dogs sit (above)."

repaired the house. Paneling from a neighboring place and bricks from an abandoned rice mill went into the mix. The result, as Gertrude explained, was not a plantation house in the traditional neo-classical vein, but "more like an English country home—a warren of small bedrooms upstairs and a few modest sitting rooms below; cozy, warm and familiar."

Eventually the acreage was expanded from 2,500 to 7,000, all of which they wished to maintain in a productive but natural state. To accomplish this, the couple took horticulture courses and looked to

The weapons above the mantle in this gun room were collected during African safaris (below). Built of logs in the 1950s, this game room held the trophies of previous decades of hunting (bottom). A lifetime of world traveling adventures are displayed in this built-in cabinet (right).

others for advice. One of these people was my father, William P. Baldwin, Jr. A U.S. Fish and Wildlife biologist, he came to Medway as a forester and wildlife consultant in 1955 and stayed on, a devoted contributor, for 30 years.

In the decade before her death, Gertrude placed two conservation easements on the property. As Virginia Beach concludes in her own history:

"[These] easements testify to Medway's legacy for the future. Not only do they articulate the significance of Medway to those who have come before ... they also verify its importance for those coming after. ... A place of such ancient and ethereal loveliness, shaped by nature's powerful forces and by people with vision, can serve as both an anchor and a transforming force in the lives of its owners, its workers and its many visitors. Its story is one of change and endurance, deeply rooted in the swamps, forests and clay soils that lie along the shore of the Back River. Gertrude's wish is that Medway will continue to provide a link between the land and human aspirations. Thanks to her foresight, that wish will be a reality."

Ownership of Medway has passed to a family member. It is on the National Register. ❧

MULBERRY PLANTATION / HOUSE CIRCA 1711, COOPER RIVER, WEST BRANCH

Certainly one of the most curious and beautiful plantation homes, Mulberry was built by Col. Thomas Broughton in about 1711. The structure has been labeled variously as Jacobean, French, and Dutch—perhaps by way of Virginia. As architect Albert Simons suggested, it seems most likely that an English architect or pattern book design underlay what was accomplished, but the the result is uniquely Lowcountry. Built of red brick in a Flemish bond, the large central element is rectangular and divided on the ground floor into four rooms and a stairwell. Usual for that early date is the absence of a central hallway. Unusual are the four small flanker buildings, which may or may not have been added for defense. These each are crowned with bell-shaped roofs of cypress shakes that complement the large jerkin-headed roof of the main structure.

A *most formal of vegetable gardens* (below).
A *garden morning. Famed Lotrell Briggs*
did the design (bottom).
Col. *Broughton's Castle at dusk* (right).

Though the house has endured its share of refurbishings, the Broughton fabric is amazingly intact. Federalist mantels and the shell-decorated corner cupboards are probably the most notable changes—and the newer addition of pine flooring that runs contrary to the first. Extra doors were apparently carved into the rear flankers and a broad dormer placed in the roof.

Col. Thomas Broughton bears particular mention. This site on a grand bluff was called Mulberry for the great trees that grew there—one tree still remains. Because it was hoped that silk would be a major export of the colony, notice had been taken and the land was acquired by founder Sir Peter Colleton. Broughton, a man of some force, built his castle here anyway. Claiming that a mistake was made, he traded 300 acres elsewhere to Colleton's son. Soon

previous spread / Clock and stair: *The Paisley tall case clock was crafted in Scotland. Built of cypress and cedar, the modest stair secluded at the rear was fairly typical of the early Georgian period (page 36, top).* **T**he cupboard: *A scalloped-edged built-in cabinet and most of the downstairs mantels, doors, and moldings were added during a 1799 Adamesque restoration of the house. Audubon's wood ducks nest nearby (page 36, bottom).* **L**iving room: *Bright sunny walls for this sunniest room of the house. As elsewhere, the furnishings are for the most part 18th century. The finest piece, a secretary bookcase with fall front desk lid and tracery doors (page 37).*

this spread / ***Library: The current owners have respected much of what they found, but their decision to strip the paint from this library woodwork adds even further to the room's comfortable feel. Note the Queen Anne-style fireside bench and a 19th-century butler tray table that doubles as a coffee table. (left).* **A *comfortable corner on a comfortable corner of the Cooper River (above, left).* **V**iew into the dining room: *A grand 18th-century, three-pedestal mahogany table with a Sheridan-style sideboard. Above the table is a William MacLeod 1856 painting of Washington, D.C, with pastoral Virginia in the foreground. In our own foreground is a walnut Chippendale chest over drawers (above, right).*

previous spread / Hidden within the trees is the new guest house (page 40, top).

"A bluff bank upon the first Indian plantacon on the right hand in the Western

branch of the North River commonly called Ye Mulberry tree." Here Broughton built

his "Mulberry Castle." (page 40, bottom) / Photograph by Alice Turner Michalak.

This forgotten elbow of the river was captured by a garden bridge (page 41, top).

A hidden gem, this cypress-shingled boathouse accompanied by

cypress knees still attached to the tree (page 41, bottom).

this spread / This recent addition was taken from Lotrell Briggs' original design (left).

"Life in the Low Country has never lacked a certain dramatic intensity of mood,"

wrote historian Sam Stoney (top). **P**erennial garden (above).

after the house was built, the Yemassee war broke out and Broughton's neighbors sought shelter at his home—an incident that suggests the house's four strategic flankers might serve a defensive purpose.

Of equal interest is Broughton's bellicose political career. Married to the daughter of the powerful Gov. Johnson, he expected to be elected to that office himself. And when an opponent's bribe brought other results, Broughton "drew together a number of armed men at his plantation and preceded to Charlestown." When they arrived, they found that the illegally elected governor had pulled up the drawbridge. A bitter parley followed, and Broughton and his men forced entry into the city. Peacemakers resolved the matter by returning the election to the colony's owners, the Lord Proprietors, who had the wisdom to select a third man for the position.

Here, too, a room with a view (below).
Despite the guest house's traditional exterior, the
two-story ceilings and pine paneling are more in
keeping with contemporary plantation planning.
French doors and windows provide a broad view of
the bending Cooper River (bottom and right).

Such was the Lowcountry shaped. We should keep this in mind when viewing the plantation homes—and most especially when we consider whether Mulberry's flankers were meant to serve a military purpose.

Like other Lowcountry plantations, Mulberry would enjoy its period of rice-growing prosperity and then, after the Civil War, be abandoned and given over to decay.

In 1914, New Jerseyites Edward Chapman and his wife purchased the property. With the aid of English architect Charles Brendon, they began an untiring effort to restore both house and grounds to their former glory. When completed, the Chapmans furnished their home with fine Charleston-made antiques—some of which remain in the house today. Since that careful restoration, Mulberry has been subject to particular care and is maintained by its present owners in mint condition.

The house is on the National Register and is a National Historic Landmark. ❧

SOUTH MULBERRY PLANTATION / House circa 1835, Cooper River, West Branch

South Mulberry has been called "the step child of Mulberry" for at times (this being one) it has been connected to Mulberry Plantation. At the beginning of the 19th century, Thomas Broughton willed to his daughter this southern half of his plantation, and the current house was built by his son-in-law, Dr. Sandford W. Barker, about 1835.

Perhaps an earlier building was only refurbished, for a massive foundation and antiquated beams suggest this. Whatever the case, the side additions came on before the Civil War, and a duck hunting club added the rear wing at the turn of the century. Though fallen into disrepair by the 1980s, the property is once more reunited with its parent Mulberry and has been carefully restored and furnished.

The present owners use it to accommodate guests and most especially to house their visiting children and grandchildren. ⚜

This gentle rise is "Harry's Hill," named for one of the last Kiawah chiefs. From here, the tribe was pushed to Kiawah Island (below). This wide wrap-around porch is a Southern classic (right).

previous spread / This turn-of-the-century bed and rocker help give the house an "Adirondack cottage" feel (page 48). **A**n antebellum parlor doubles as a game room. Above the mantel is a Currier and Ives print, "Game Fish." On the mantel, there's a playful collection of shore birds and fir trees (page 49, top). **F**or a house so distinctly different from their Mulberry home, the owners felt they could relax a bit and trust themselves to a decorator and to things that had come their way. This dining room is a result. The pine floor is painted in a tile pattern (page 49, bottom).

this spread / An accomplished naturalist, Dr. Barker would often invite friends for "botanizing" weekends (top). **A**n aerial view shows the relationship of South Mulberry to Mulberry Plantation. In the distance is a well-disguised industrial park. / Photograph by Alice Turner Michalak (above). **A**rchitect Glen Keyes designed this mahogany greenhouse. Mullioned windows and graying posts and beams make a fine backdrop for orchids and amaryllis (right).

DRAYTON HALL PLANTATION / HOUSE CIRCA 1738, ASHLEY RIVER

John Drayton began his house in 1738 and finished it four years later. "Mr. Drayton's Palace on the Ashley" was a description of that time, and still today Drayton Hall is addressed with superlatives. Many people believe that it's the finest example of Georgian Palladian architecture in America. Being influenced by early Roman building, Italian architect Andrea Palladio used great columned porticos to front his domestic architecture. This style was quite popular in England. Here on the raw American frontier, there's no known architect. However, one of the massive Georgian overmantels has been linked to a pattern book. Master builders worked from these books, and it is possible that the owner hired just such an accomplished carpenter.

Nevertheless, Drayton must have given whomever was responsible a difficult time, for the front entry was widened at the last moment with no regard to the load-bearing foundation beneath.

Held in trust, Drayton Hall is opened to the public daily. Much of the tour guide's time is spent dispelling the illusions of how life was actually lived inside an 18th-century Georgian plantation house as opposed to the Hollywood version of the Southern plantation. The reality had far more attachment to English architecture and manners.

The entry hall was a public room that housed the stairway leading up to a second-story ballroom. Guests were entertained there, and the residents would withdraw to side rooms for privacy. Over the years, the use of the rooms naturally changed. At first, furniture was sparse and pushed against the wall, but gradually it moved to a more comfortable center of the room. And dining was rather haphazard, no great banquet tables, only many small ones.

The ceilings were reported at the time to be unfashionably low. The first-floor ceiling was only 12 feet high, and the second a mere 14. One drawing-room ceiling boasts wet plaster carving—the only surviving in America. The

Drayton Hall is considered by many to be "the finest early-Georgian house in America" (below). This overmantel in the first-floor great hall appears to have come from William Kent's Designs of Inigo Jones. *The animal in the crown of the broken pediment is either a fox or a wild boar, depending on the viewer's imagination (right).*

Entering from the river side, the guest was greeted by this soaring, mahogany-embellished stairway (left).

This drawing room boasts one of the finest ceilings in America. The design was carved into wet plaster. The result is a delicate masterpiece that, unlike many of the mansion's ceilings, has managed to survive (above).

previous spread / Federalist windows replaced the originals after an early hurricane.
The thick walls made room for the easy convenience of indoor shutters. These are still used today to
control the light entering the room and to protect in time of storms (page 56, top). **T**he walls were first
painted a sand color and the trim accented by two darker shades of the same. But in the
Victorian era, most of the rooms received this treatment of blue. The rosettes of the wall border
are repeated in star studded splendor above (page 56, bottom; page 57).
this spread / Rather restrained pilasters of the Doric order encase this opening into the stairwell.
On the far side was the simplest of the house's moldings, but elsewhere the complexity of orders only
increased (left). **A** majestic essay in mid-18th century ornamental brickwork (above).
following spread / These two Adamesque mantles were installed in 1803. Sadly, a third
and the most beautiful was ripped out by art thieves in the 1960s (pages 60–61).

interior decor, in general, can be described as massive, balanced, and elaborate. The intricately carved decoration has had only one coat of paint since it was built. Indeed, relatively few changes have been made anywhere. Federalist windows went in when the first blew out in a hurricane. Some Adamesque mantels, Victorian wood ceilings on the second floor, and shingles on the roof pediment were added later. Mrs. Charlotte Drayton lived here through much of the 20th century without electricity or plumbing, so there were no such modern intrusions to remove in search of authenticity.

Drayton Hall was once as famous for its garden as for its mansion, but not many of the plantings remain. Once 30 support buildings stood here. All are gone now except a solid little brick privy. In the front we see the brick foundations of the two flankers, one a kitchen, the other perhaps a laundry or bachelor's quarters.

It was said of John Drayton, "Such was his character, he lived in riches, but without public esteem. He died in a tavern, but without public commiseration." Upon his death, his fourth wife, the 16-year-old Rebecca, used the estate to provide for six of her slaves. She retired to Charles Town, lived to be over 80, and left what remained to her grandchildren.

The house survives today because the resident Drayton convinced Union troops that it was a smallpox hospital. Following the Civil War, phosphate mining allowed it to stay in the family's hands.

Perseverance and luck have kept Drayton Hall standing over the years. The mansion is far more fragile than its stately appearance suggests, and the National Trust is simply trying to maintain an authentic treasure. That alone is expensive business and it depends on public support.

The house is on the National Register and is a National Historic landmark. ⚜

Sun-painted shadows—a lyric poem (below). *Drayton Hall: "Its very existence proves its strength against the tests of time," writes the National Trust* (bottom). / *Photograph by Alice Turner Michalak*
The flagstones cover an earlier brick floor in this doctor's office (right).

MIDDLETON GARDENS / House circa 1741, Ashley River

The first Henry Middleton gained this property in 1741 as part of his wife's dowry. He added two flankers to an existing building and began an extensive formal garden.

It took 100 slaves almost a decade to complete the terraces, walks, lakes, and vistas that are still enjoyed today.

Henry was president of the First Continental Congress, and his son Arthur Middleton signed the Declaration of Independence. His son Henry Middleton (who was governor of the state and minister to Russia) was an avid botanist. Assisted by Frenchman André Michaux, Henry planted thousands of exotic and domestic plants and added much to the garden.

Middleton's surviving flanker makes a perfect house museum (below).

"Among the specimens of art in this elegant collection," wrote an 1840 visitor, "we observed a piece of statuary, the counterfeit presentment of a young female, of such exquisite softness and delicacy of finish, of such perfection of feminine beauty, as to give it a very exalted rank among the achievements of the chisel" (right).

His son William may have made the single-most important contribution to the garden—the azalea. William is also remembered as a signer of the Ordinance of Secession. After the war, he returned to find himself the custodian of ruins.

The early home was burned by Union troops, and the earthquake of 1886 brought down all of the walls except those of one flanker. And it's this "bachelor quarters" flanker that's been restored in a slightly romantic Jacobean style and now serves as a museum house.

What remains of the home is operated by a foundation as a public trust.

The members of the Middleton family have been particularly generous in donating fine heirlooms of true museum quality. Hence it was possible to re-create at least a portion of what was there and give the visitor a solid sense of the splendor now passed.

Middleton Place is on the National Register and a National Historic Landmark.

previous spread / "T_he formal terraces and lakes at varied levels may recall better known_
European gardens," wrote E.T.H. Shaffer, "but the live oaks and Spanish moss weave a spell
that is of a New World and an Old South" (page 66, top) / _Photograph by Alice Turner Michalak._
T_his graceful Oriental crane joins the exotic azaleas and camellias_
that arrived here long before (page 66, bottom). **A** _rustic arch invites a detour_
from the garden's more formal byways (page 67, top). **T**_ulips, pansies, and more corralled_
by boxwood. In the far distance, a restored mill house (page 67, bottom).
this spread / A _flanker building_ (left). **A** _riotous celebration of color._
Azaleas bursting into bloom (top). **W**_hite swans adrift on black water._
Spring at Middleton Place (above).

*M*usic room: In 1820, Henry Middleton served as minister to Russia and returned home
with this playful French painting of the reluctant listener. And the gold tiara on the table
was worn by Mary Helen Middleton at that imperial court. An even grander reminder,
however, is the tremendous "Picture Gallery at the Winter Palace"—a painting that
seems to expand the room. The 1815 music stand is mahogany and English made.
Note the candle holders, a slightly later American addition. The marble-top tables, sofas,
and chairs all date from about 1820 and are probably Philadelphia made (above, left).

*C*hild's bedroom: The child's bed and slant-top desk are American made, walnut,
and date from the mid 1800s. The toy bed is as old. It is mahogany and English made.
The remainder of the toys are 19th century and American in origin. But such a recital
hardly does justice to the pleasure they once brought to their young owners or the
pleasure they now bring to modern visitors of all ages (above, right).

*M*aster Bedroom: Upon a Charleston-made "rice bed" lies the actual clothing of
Henry Middleton, the 1741 founder, and Arthur Middleton, the signer of the Declaration of
Independence. The handsome trunk in which these items were discovered is
of leather-trimmed camphor wood (right).

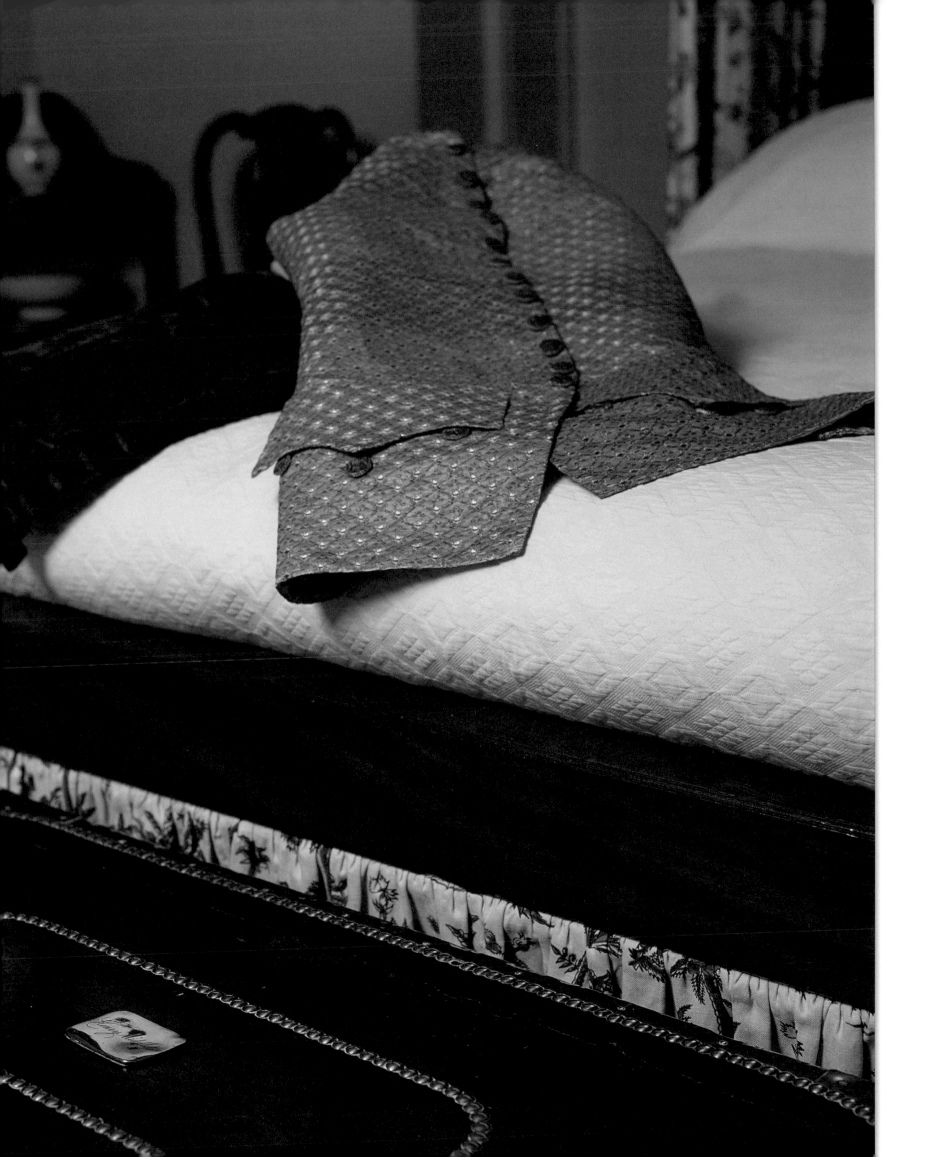

MAGNOLIA GARDENS / HOUSE CIRCA 1873, ASHLEY RIVER

John Drayton (the father of Drayton Hall builder John) arrived here from Barbados in 1679. His own mansion burned, and its replacement was burned by Union soldiers. The current house dates in part from before the Revolution and was the family's hunting lodge. Taken apart and floated 15 miles down the river in 1873, it was placed on the previous foundation. The Victorian details were added then and the distinctive tower for artesian water. The encircling columns were added within the last decade.

Today's plantation garden spreads out from the visitor's center in a happy informality—and being an informal garden is what the designer John Grimke-Drayton had in mind. A hunting accident took his brother and left the young theology student heir to the family fortune. He completed his degree, married his Philadelphian bride, and settled here to tend to the needs of his flowers and those of his flock in nearby St. Andrew's church.

The garden that he grew is laid out in the Romantic pattern, which was then popular in England. Before this time, geometry and strict pattern had been the rule. But now a more personal expression was allowed and greater value was placed on the existing forest. In 1843, he imported Camellia japonica, and five years later Azalea indica. In 1851, tuberculosis caused him to retire.

The Civil War brought ruin, and in the aftermath, Grimke-Drayton was forced to sell much of his property and lease the rest to a phosphate company. He hated this stripping of the soil (in his will he forbade its happening again), but still he managed to save his garden and opened it to the public in 1870. Visitors came on paddle-wheel steamboats. Baedeker's travel guide listed it with only two other American attractions—the Grand Canyon and Niagara Falls.

The furnished house is open to tour, and visitors can find a variety of botanical wonders down each curving path.

The house is on the National Register. 🔺

*this spread / **A** happy collection of styles—the house has at its core a modest pre-Revolutionary dwelling (below). **C**olumns (right).*

*following spread / **"S**olid masses of blossoms in all the shades of red from palest pink to the deepest crimson, and then a pure white bush," wrote an 1875 visitor (pages 74–75).*

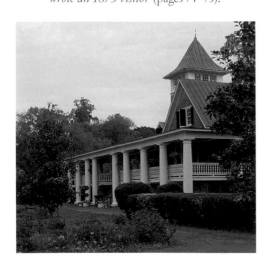

previous spread / **A**n *eclectic collection reflects the broad interests of the Rev. Grimke-Drayton and the family members who would follow* (page 76, top). **D***ining room: The china is Chinese export. The family began with 600 pieces of this Lostar pattern, but through the vicissitudes of a turbulent existence, only 81 remain. The service is of American coin silver* (page 76, middle).

S*tudy: One of the house's finest pieces, a Chippendale slant-top secretary bookcase, and in accompaniment, a Queen Anne-style side chair. Most of the furnishings date from the house's antebellum occupancy and many are family pieces* (page 76, bottom). **D***esk: ... as if the venerated priest and gardener had only stepped away for a moment* (page 77).

this spread / **A** *youthful David* (left). **A** *cat on a not-so-hot asphalt roof* (top). *The Drayton Family Tomb dates from before 1700. The carving was by Jardella, the nation's first noteworthy sculptor. Union soldiers left behind the marks of saber and ball, and the 1886 earthquake cracked the vault* (above).

previous spread / L*ady Banksia roses* (page 80, top). **S***naking rails beneath wisping moss.
A common sight in the early "open range" days* (page 80, bottom). **T***he Union Army and
accompanying government agents made short work of the Ashley River plantations statuary.
Piece by piece, it's being replaced* (page 81, top). **A** *hallmark of these gardens—the ornate white
arch of bridge across a cypress-filled lagoon* (page 81, bottom).

this spread / "I *have seen gardens, many gardens, in England, in France, in Italy,"
wrote novelist Owen Wister, "but no horticulture that I have seen devised by
mortal man approaches the unearthly enchantment of the azaleas
at Magnolia." / Photograph by Alice Turner Michalak* (above).

A *knot garden: boxwood and euonymus* (right).

BOONE HALL PLANTATION / House circa 1936, Wapeckercoon Creek

Among the first of the colonists, Major John Boone had established his plantation by 1681. An Indian trader, he was closely linked with the resident Sewee tribe, and his purchase of Indian slaves from them and his dealings with pirates got him expelled from the colonial assembly three times. But many of his neighbors did the same without being prosecuted, and his expulsions were probably rooted in political quarreling. His transition to a more staid and responsible citizenship was rapid.

John's son Thomas was a distinguished servant of state and church. He died in 1749 and may be buried here beside the avenue of oaks that he had planted. John's daughter was the grandmother of two of South Carolina's most noted statesmen, John and Edward Rutledge.

In 1817, the Horlbecks, a family of noted builders and engineers, bought the property. They planted cotton and ran a large brickworks. The entrance avenue was enlarged at that time, and the slave cabins were built. The Plantation's main house burned in the last century, and in 1936 the present Georgian mansion was constructed with the previously used bricks.

*this spread / **T**he finely worked wrought-iron gate complements the house* (below). ***A** classic avenue of oaks* (right). **following spread / **P**icture perfect.* *Perhaps the most photographed plantation house in America. It's easy to see why* (pages 86–87).

The owners reside in this house, but much of the downstairs is open to tours and those are the rooms pictured here.

An amazing 15,000 pecan trees were planted in the 1890s, but these aging trees are passing on. Now, a century later, "you pick it" strawberries and vegetables have found an equally enthusiastic market.

Boone Hall is opened to the public daily. Fine horses graze in the pasture, and polo is played in the fall. Major battles of the Civil War are fought by enthusiastic re-enactors. And moviemakers have found it to be an ideal location. Boone Hall is truly a "working plantation," what has been called "a 21st-century plantation."

The plantation slave quarters are on the National Register. 🌿

this spread / A *brick-lined patio allows the sunroom to extend into the sun* (above). **W***hite wicker dating from the house's construction gives the proper relaxed atmosphere to this sunny room* (right).

following spread / **T***he table in the dining room is Hepplewhite and dates from the late 1700s. Sheffield silver plate and Royal crown derby china are displayed on the table. All this and crystal, too, reflected in an 11-foot-tall, free-standing Empire mirror* (page 90, top). **T***his graceful cantilevered stair rises from or perhaps sweeps down upon a teakwood parquet floor. The short case or "grandmother's" clock is English made and a century old. To the right is a marble-topped pier table. Throughout the house, electrified chandeliers and oil lamps suggest an earlier time* (page 90, bottom). **T***he family has some heirlooms dating to the 1800s, and they have added other suitable pieces. Beside the fireplace is a tilt-top pie crust table from the late 1700s. The piano is a century older. The sofa and chair are Italian made of intricately carved olive wood. The painting is "Queen of May" by Kinsman* (page 91).

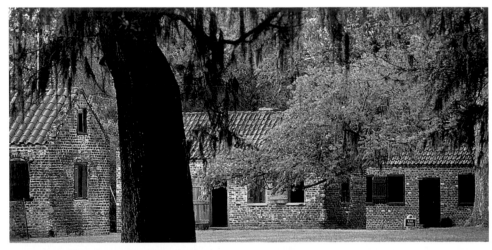

previous spread / "Two paths diverged and I ..." For those who choose either path,

there are grand rewards (page 92, top). A profusion of blooms. Master gardener Ruth Knopf explains,

"I use purple for my neutral color, it blends and separates the other color beds" (page 92, bottom).

These formal gardens that front the house have been recently restored, and the roses,

in particular, are responding "bloomingly" (page 93, top). The collection of antique roses

makes this garden "a museum of sorts" (page 93, bottom).

this spread / These sturdy brick slave quarters probably housed the house servants.

They've recently been restored and are on the National Register (top). A herringbone path of

ancient bricks (above, left). A bouquet of flower, leaf and stem (above, right). / Photograph by Alice

Turner Michalak. The antebellum gin house is now home to a restaurant and gift shop (right).

DEWEES ISLAND / Houses circa 1990s, The Atlantic Ocean

In the forefront of America's movement toward a "green" or environmentally friendly architecture is the new Dewees Island community, and through it, lessons in a sustainable building and living process are being learned and disseminated.

Fittingly, Dewees and neighboring Capers Island escaped agricultural development in the first years of settlement, remaining on the map only as the "hunting islands" of the Indians. Eventually wild cattle roamed here and live oakers took oak limbs and cedar trees for ship building.

The planting of sea-island cotton followed the Revolution, but it's not certain just how successful this was—on the sand beach ridge, you're especially vulnerable to hurricane destruction.

Fitting, then, that the island has found a use for which its Florida maritime forest and several miles of beach shore are amply suited. John L. Knott, Jr. has written a Dewees philosophy that explains exactly what the problems facing us are, and since they apply most particularly to the still undeveloped plantation lands, we'll repeat a portion:

"The philosophy of Dewees is that we are an essential participant in the environment. We simply need to re-discover our intuitive base as to how to live in harmony with our environment as well as relearn how to select habitat and nest within it (as opposed to dominating and destroying more than we need). The history of great architecture and historic communities throughout the world is a history of designing and building in context of natural elements and geography. With that in mind, the question arises, 'What changed our ability to plan design, and construct communities and buildings that functioned so well and felt so good?'

"As the industrial revolution evolved and our

*this spread / **A**n osprey nest and the question it raises—"Is man a functional part of the natural system or an interloper that has no place on this earth?" (below). **H**ouses are built well back from the beach, leaving a beauty to be shared by all (right).*
*following spread / **T**he island offers a variety of habitats, from oceanfront to freshwater impoundments and all that's in between (pages 98–99).*

technological prowess emerged, three major forces altered our thinking. The first was a sense of control and power that allowed us to do anything we wanted. Electric power, air conditioning, elevators gave us a sense that we could build anything, anywhere, anyhow because we could control what was inside. The second was telecommunications, which eliminated the need for proximity to afford effective communication. The third was our automobile passion, which has shifted our focus from planning for people to planning for cars ... which also had the net effect of destroying public transportation systems, isolating communities and de-personalizing our lives. Add to these factors, a sense that all resources were unlimited and you have the recipe for our current crisis; a crisis where people and the environment have become enemies."

In fact, the antebellum planters felt themselves to be a part of the Industrial Revolution that was only just beginning. Fortunately, that modernizing revolution has passed by, and like Dewees, many have been maintained over the last century in an untroubled isolation.

While plantation architecture might not have been in complete accordance with mother nature and at times the hunters were perhaps too "avid," in general stewardship, at least, Lowcountry plantations have been "green" for quite a while.

For Knott the "purpose" of Dewees is to extend these principles to the entire country by constructing a community where environment and development are natural allies, a place where technology does not dominate and where such values are taught to the point they become "intuitive."

The result, in part, are houses similar in construction to the Huyler House. 🐾

This dune fence captures not only the sand, but a bit of island beauty (below).
Boardwalks offer views, breezes, and protection for the vegetation beneath (bottom).
The Huyler House was inspired by the earlier beach houses of Sullivans Island and Pawleys Island. This great room offers a spot for receptions (right).

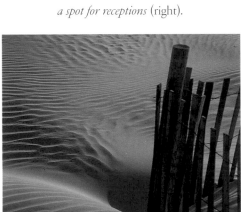

By 1690, French Huguenots had settled on the South Santee River and a generation later Baptist leader George Screven founded Georgetown. In 1732, Georgetown was designated as an official port, and being at the confluence of six rivers, it's not surprising that as a port she prospered, especially when it was discovered that these same rivers could be used to irrigate rice fields. By 1840, the Georgetown district grew almost half of all the rice produced in the United States. And by historian George Rogers' estimate, this was being done by only 91 planters.

Georgetown and Her Rivers

HAMPTON PLANTATION / House circa 1730, South Santee River

The earliest portion of the Hampton Plantation house was built by a member of the Horry family sometime between 1730 and 1750. French Huguenots had settled here in the Santee wilderness in the preceding century. The home was sturdy and convenient Georgian architecture, saltbox shaped with small-paned windows, low ceilings and wood-paneled walls. Two large wings, one containing a ballroom, and a tremendous neo-classical portico were added sometime before 1785 by Daniel Horry and his wife Harriot, with Harriot's mother Eliza Lucas influencing the design.

While living in London, Eliza Lucas Pinckney and her daughter Harriot had often visited Shakespearian actor David Garrick. Returning to the Lowcountry, they borrowed from his country estate both the name Hampton and this neoclassical portico design. Harriot's husband had used the expanse of the front lawn as a race track. One oak by the door had been spared, however. General Washington spoke on behalf of the survivor and it remains today the "Washington Oak" (below).

This ballroom was probably added just before the Revolution. Each piece of pine flooring runs the length of the room. Each section of the wide paneling is from a single piece of cypress (right).

When George Washington passed this way on his Southern tour in 1792, the now widowed Harriot and her mother met the president wearing sashes painted with his likeness.

The plantation next passed through marriage to the Rutledges. This family is best known for their extensive service during the Revolution, and in more recent times for the literary works of the state poet laureate, Archibald Rutledge.

Archibald had grown up on the plantation at the end of the 19th century, then spent his adult years teaching in Pennsylvania.

In 1936, he retired to the now-abandoned Hampton Plantation and set about refurbishing both the house and grounds. In the process, he wrote perhaps his best-known work, *Home by the River*.

Having little money at his disposal, this restoration was fueled by love and what today is called "sweat equity." Though the scattered live oaks are fittingly ancient, the holly and dog-

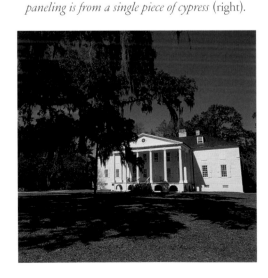

1787 *Inventory of Daniel Horry's ¶ Though the mansion house at Hampton is now empty, we may use this 1787 inventory of the Hampton furnishings and imagine and interior. The values given are in pounds. The list was made following the death of Daniel Horry. ¶* **Long Room** *(Ball Room): 1pr. Guilt Framed Looking Glasses £25, 1pr. Girardoes £20, 2 Sofas £12, 8 French arm chairs £20*

1 Set Brss Chimney Furniture £3, 2 Card Tables £4; **Blue Parlor**: 1 Harpsicord £60, 1 Guitar £5, 1 Music Case £2, 2 Card Tables, 7 Prints £7.10, 1 Guilt Frame Looking Glass £6, 9 Mahogany Chairs £3, 1 Carpet £7, 1 Set of Brass Chimney Furniture £10, 1 Square Table £4, China in Closet £20; **Hall** (Entry): 1 Mahogany Bookcase £8, 1 Marble Slab £5, 2 Mahogany Tables, 1 square, 1 round £3.10, 9 Chairs, 1 Carpet, 1 Matt, 1 Japaned Plate Warmer £15.1, 2 White Framed Looking Glasses £8, 190 Volumes of Books £38, Sundry Articles in Closet £1; **Dining Parlor**: 1 Large Mahogany Slab £5, 2 Square Tables £6, 1 Tea Table £2, 1 Set of Nankeen China £3, 1 Large Turkey Carpet & 1 Matt, 1 Sofa, 14 Mahogany Chairs, 8 Prints, 1 Set of Brass Chimney Furniture, 2 Glasses, Gilt Frame Total for all £65; **Bed Chamber Below**: 1 Bedstead with Bedding & curtains £40, 1 Double Chest of Drawers £8, 1 Single Chest of Drawers £4, 1 Pr. & 1 dressing glass, 5 Mahogany Chairs £8, 1 Set of Steel Chimney Furniture, 1 Bason Stand £1.10; **Study**: 1 Mahogany Desk £3, 1 Press £1, 1 Mahogany Chest, 200 books, 1 small Bedstead, bedding & curtains, Closet £45; **Passage**: Mahogany Cased Clock £12, 1 Square Mahogany dining Table £2.10; **Blue Bed Chamber**: 1 mahogany Bedstead with Curtains & Bedding £50, 1 Toilette Table & Toilette Glass, 6 Mahogany Chairs, 1 Wilton Carpet, 1 Set Steel Chimney Furniture £13; **Big Room**: 1 Mahogany Bedstead with Curtains & Compleat £60, 1-1/2 Chest of Drawers, 1 Dressing Glass £5, 7 Mahogany Chairs, 1 Wilton Carpet £18, 1 Set Steel Chimney Furniture £.10; **Little Room**: 1 Mahogany Bedstead with curtains, etc. £35, 1 Small 1/2 Chest of Drawers, 6 Mahogany Chairs, Set of Chimney Furniture, 1 Dressing Glass £6.1; **Middle Room**: 1 Mahogany Bedstead, PB Curtains & Bd Compleat £15, 1 Dressing Table & Chimney Glass, Mahogany chairs, 1 Roll of Wilton Carpet £14; **No. 12**: 2 Mahogany Bedstead with curtains & Compleat £40, 1-1/2 Chest of Drawers, 1 Dressing Glass £6, 1 Carpet, 8 Mahogany chairs £5; **Pantry**: articles of China, etc. £10; **Garrett**: PB Trunks, PB Furniture £15, Kitchen Furniture, Saddles, Bridles, & other Horse Furniture £22.47. Under Stock was listed 44 oxen, 116 "stock" cattle, 77 sheep, and 84 hogs. And at Hampton proper were the following horses: Grasshopper, a pair of sorrels, the gray colt Duke, the bay colt Pilgrim, Little Dick, Leeds, the black gelding Lofty, Swap, Gray Betty, and Cream Pot.

Harriot Horry chose 26 recipes of her mother Eliza Lucas' to include in her own cookbook. Richard Hooker gives this list in *A Colonial Plantation's Cookbook*: To Pott Beef like Venison, Black Caps, To Dobe a Rump of Beef, Plum Marmalade, Beef Collops, Orange Marmalade, Stewed Mushrooms, Almond Cream, To Caveach Fish, Coco Nut Puffs, Snow Cream, Cheese Cakes, Baked Pares, Egg Pyes, Yam puddings, Orange Pudding, Little Puddings, Rennet, Mince Pyes, Pickle for Hams, Tongues, Rusks or Dutch Beef, Mushroom Catchup and powder, Ratifia, Queen Sauce, Orange Flower Ratifia, Duke of Norfolk Punch. And beside these delights for the Horry table, Daniel's wife offered almost a hundred more recipes of her own, as well as offering advise on washing out grease or paint, fattening oxen, making cheap paint, preserving cream, washing silk stockings, dying a good olive brown, dying pink, making a strong cement and much more.

The restoration and expansion of the plantation garden was Dr. Rutledge's ongoing delight (right). "Married in 1768" ... "Ditto Ditto 1789" Since her mother had married and moved here in 1768, the playful notation at the bottom was probably made by daughter Harriot who eloped with Frederick Rutledge in 1789 (page 108, top). Cat in a window — molten glass spread seamlessly between two cylinders (page 108, bottom). Seen through the ballroom's doorway is the house's first parlor. The finish of this pre-1750 room is cypress paneling grained in a somewhat primitive fashion to look like mahogany (page 109).

wood avenues and much of the 6-acre camellia garden date from this period.

In 1971, the state acquired Hampton Plantation for use as a park. At that time, the house was given an extensive and more conventional restoration with particular emphasis placed on preserving the house and exhibiting its structural components. Still, romance aplenty haunts those halls and surrounding woodlands, and in the fall, winter, and early spring, no public place in the Lowcountry is more serene and beautiful.

In the course of expansion, this set of steps probably changed directions. The house's first main entry was here toward the river. Then with the building of a "river road," canoes and flats gave way to horse and carriage, and guests were invited in the back—to which a tremendous portico had been added (below).

This exceptionally detailed mantel was probably constructed in Charleston and shipped up the Santee. The Adamesque details are centered on a neoclassical urn from which a spiraling of Acanthus vines, flowers, and leaves cascade. For the Greeks, this motif was associated with fertility and wine. It's not known for certain what significance it held for the Hampton residents of the 18th century (right).

Here's Archibald Rutledge's version of the story:

"The days at Hampton are full of friends, of work, of hope. All about is a sense of calm, of joyous relaxation. The quail call, and the doves from the oaks; the wide fields sleep in the sunlight. There is here a gentleness that suggests what life can be; that in the realm of the heart, inglorious is the victory of might. There is something that persuades one to accept quietly life's stillness and its song. Only by such reconcilement can we find an answer to life's longing. Here, in this deep harbor of eternal dreams, the spirit finds sanctuary and is able, untormented by life's badgerings, to meditate on the mystery all about us. The calm of the natural world must be the peace of God."

Nothing of substance has changed since that was written 60 years ago.

The house is on the National Register and listed as a National Historic Landmark. ≈

SILVER HILL PLANTATION / House circa 1791, Sampit River

"**T**his was the country of yellow sand and golden sunshine, of the Magnolia grandiflora and the blood-red Maple, of the Cherokee Rose ... of the yellow Jasmine ... of the tall Turpentine or Yellow Pine and the wide-spreading Live Oak ... and the Sweet Olive that carries you with one breath right up to the gates of Heaven." So wrote Elizabeth Withers Forster of her childhood on neighboring Friendfield Plantation. Silver Hill was by that time what some refer to as "the world's greatest haunted house." Young Elizabeth and her friends played there and left behind a volume of teasing graffiti.

"This mystery house" is restored now to mint condition (below). House, rice field, and the bending Sampit River (bottom) / Photograph by Alice Turner Michalak. Morning sun brings a touch of gold to both the marshes and this best room in the house (right).

The Silver Hill house was built around 1791. But as records were lost during the Civil War, it remains for the rest "a mystery house," a structure so forgotten that the present owner had difficulty obtaining a building permit. Apparently no taxes had been paid for the last 150 years, and Silver Hill simply didn't exist. Still, he prevailed.

The recent preservation effort was a detective story in itself. The foundation for the completely missing porch was found below grade, and what went where was painstakingly resolved. The stairs were reimagined, and for the first time ever the conveniences of plumbing and electricity were added.

When the original marbleized and grained finishes deteriorated, they were photographed, shellacked, and repainted in a similar manner. Tom Grahamn, an old friend of mine, was the contractor, and my nephew Chris Bates and my son Aaron did much of this work. ༄

previous spread / A *handsome landside entry. Note the irregular spacing of windows and dormers. Formal gardens on each side add a touch of symmetry (page 114).* **L***andscape architect Jean Rothrock filled these formal beds with cabbage, kale, rosemary, and mint and bordered all with boxwood (page 115, top).* **B***rickwork and white pickets anchor the house. Traditionally, plantings did not go directly against the building (page 115, middle).* **C***amellias, azaleas, and seasonal plantings create the knot in this simplified knot garden (page 115, bottom).*

this spread / T*he chest in the center hall is 18th century and South Carolina made. The print is Audubon's Rosy Spoonbill (left).* **A** *masterful restoration. That missing front door was found in a nearby shed stacked with paint cans and serving as a workshop table. The faux mahogany with gilded trim is a re-creation of what was there (above, left).* **I***n the 1930s, the complex dentil crown molding had been stripped for use in the house next door. That could be copied. The mantel, however, was altogether gone and had to be painfully reconstructed with just the use of one faded photograph. The medallion at its center celebrates a sheaf of rice. On the wall is a convex mirror of the Regency style and an 1852 painting on glass—a bark under full sail (above, right).*

previous spread / **T***he 18th-century English furniture, if not of the house, is at least*
contemporary to it. The mahogany table is George II (page 118). **T***his handsome chandelier is*
a reproduction—but one made in Venice by the same company using their same 1790s patterns.
The owner went there to make design choices, and when the carefully packed crystals arrived,
he had the pleasure of assembling them himself (page 119, top). **A** *Venetian sconce lit by candles.*
Note the delicate snuffer beside each glass chimney (page 119, bottom).

this spread / **I***n the late 19th century, children and adults left behind their*
pencil-scribbled thoughts. Playful Frances Elizabeth Forster declared herself a "spinster."
She was 17 (above). **I***n order to preserve the building's fabric, it was necessary to tuck the*
plumbing in as unobtrusively as possible. A washstand was provided for each bedroom,
and wardrobes took the place of closets (right).

ARCADIA PLANTATION / HOUSE CIRCA 1794, WACCAMAW RIVER

Dr. Isaac E. Emerson bought the Allston plantation of Prospect Hill in 1906. He added neighboring Clifton and four other places and rechristened the whole "Arcadia." Thomas Allston had begun the construction of the Prospect Hill house in 1794, but it fell to his widow to complete the "house frame," which she did with handsome results. The double-tiered portico with Ionic columns and hipped roof suggests a refined Federalist version of Charleston's famed Miles Brewton house, which was in the Allston family at that time. Also of note are the double brick entry steps and faux painted window

When President Monroe visited, a carpet was rolled out from the mansion all the way to the river—200 yards. But as Alberta Lachicotte suggests, "Possibly the carpet has been extended a few feet with each telling of the tale" (below).

This large octagonal addition has a turn-of-the-century mahogany table constructed of 52-inch planks. The chairs are Centennial (1870s) Chippendale style. The portrait above the mantel is of a young Cornelius Vanderbilt—the present owner's father (right).

of the portico. Thomas Allston's brother William built an identical house close by at Clifton that burned soon after completion.

Dr. Emerson did much to return Prospect Hill to its former glory—rescuing the house, refurbishing the terraced gardens, and for the sake of comfort, refiguring some of the house's interior and adding a handsome wing to each side. Dr. Emerson's daughter married a Vanderbilt, and their descendants continue to live here.

The house is on the National Register.

previous spread / **T**he landing's grand Federalist window was once a door leading out onto
the roof of the land-side entry. Now the family's protectors keep watch from down
below (page 124, top). **T**his bright, tile-floored breakfast room is a relatively new
addition (page 124, bottom). **A** step down sideboard is flanked by elephant tusks.
Racing along a riff pine and mahogany floor is the owner's father's toy
cart and horse—real horse hair covers the horse (page 125).
this spread / **T**he removal of two walls created this spacious living area and doubled the
complement of fine dentil cornice, paneled wainscoting and handsome Adam-style mantel (left).
This hall was once a portion of the Allston family's second-story ballroom. Within the dollhouse
is a remarkable collection of finely detailed furniture. And with its fine view of the terraced gardens
and distant Waccamaw River, the second-story balcony no doubt offered a retreat (above).

previous spread / **A** *view of the Arcadia house and Dr. Emerson's famed*
"Arcadia gardens" (page 128, top). **P***arallel arches. White pickets reflected in*
black water (page 128, bottom). **A** *classic combination: live oaks and camellias* (page 129, top).
A *traditional English carriage house complete with carriages. One accomplished ancestor was*
Queens Master of the Horse. He drove the London to Brighton coach three days a week with
such ability that a monument to him is on the Brighton Road (page 129, bottom).
this spread / **A***n early example of the handsome Federalist double-tiered portico* (left).
A *garden path* (above, left). **T***erraces, pools, gateways—and on all sides a host of azaleas*
and camellias. Note the star-shaped reflection pool at the center of the
lower terrace (above, right) / *Photograph by Alice Turner Michalak.*

DOVER PLANTATION / House circa Early 19th Century, Winyah Bay

Dover was originally a part of Winyah Barony, a 12,000-acre tract granted on June 18, 1711, to Landgrave Robert Daniel. One day later, this property combined with other acreage was transferred to Landgrave Thomas Smith. In 1756, two of Thomas Smith's sons, Henry and Benjamin Smith, sold their land to Elias Horry. This tract passed later to his nephew, General Peter Horry of Revolutionary War fame, who formed his three plantations: Prospect Hill, Dover, and later Belle Isle. He named Belle Isle in honor of his good friend Francis Marion's birthplace.

The first house at Dover Plantation burned. The main house that is there now was at first a residence for the cotton plantation, Woodlawn, in Berkeley County. Woodlawn now lies beneath the waters of the Santee-Cooper Lake. The house, however, was bought and saved by Mrs. Henry M. Sage of Albany, New York. Carefully dismantled, the structure was stored and later re-erected by Mrs. Sage on the Dover property in 1949. It became her winter home for the next 20 years and at her death was left to her son, and later to his four children.

*"**W**aters of Santee-Cooper dam have submerged the lands of Woodlawn," wrote Alberta Lachicotte, "but the plantation lives on in its proud mansion that stands on Winyah Bay" (below). Photograph by Alice Turner Michalak*

***T**his handsome country house entry was borrowed from a Savannah townhouse that was about to be demolished (right).*

In 1995, the Sage family sold the house to the present owners, who make it their principal residence. The couple restored control to all but one of the original rice fields and purchased additional acreage to bring Dover back to 1,000 acres.

Together with their designer, David Meroney, the owners have restored, furnished, and put back into use the guest cottages, stables, and other buildings. The formal gardens and grounds have been restored. The main house has been lovingly restored and furnished and once again is back to Alberta Morel Lachicotte's description: "Despite immense spaciousness it conveys an atmosphere of warmth and softness not always achieved in large dwellings. Many plantation mansions of today, though imposingly styled, have a cool aloofness successfully avoided here." ⚓

previous spread / Dover reflected (page 134, top). **T**his harbor entry doubles
as the Intracoastal Waterway, and the far shore is, or is close to, the site for the first
European settlement in North America (page 134, bottom). **S**unrise (page 135, top).
Pavilion columns frame a perfect moment (page 135, bottom).
this spread / Roses are the owners' pride and joy. This open pavilion encloses light, texture,
and even aroma (left). **A** Georgian cottage. Careful restoration was extended
to the outbuildings (top). **G**rape vines—such arbors were once a plantation must (above).

previous spread / *T*he *exceedingly rare wallpaper was hand-blocked by Zuber and Fils in 1827.*
The Chinese porcelain vase is from the Ming Dynasty. The 18th-century mahogany breakfront
bookcase contains Chinese porcelain of a slightly later period (page 138, top).

*D*ark *leather and even darker paneling lend a rich and comfortable air to this gun room. Above the*
mantel is a portrait by Carl Guther of General Lee and his horse Traveler (page 138, bottom).

*T*his *hall addition was borrowed from another house and the stairway from yet another.*
(Both fated to be destroyed.) The Queen Anne double bonnet secretary with original
chinoiserie dates from the 18th century (page 139).

this spread / *T*he *silver is part of a collection of American coin silver* (above).
*A*n *American-made Sheraton-style table and Hepplewhite chairs and an English Regency chandelier.*
The portrait is by Joy Thomas (right).

CHICORA WOOD PLANTATION / HOUSE CIRCA BEFORE 1819, PEE DEE RIVER

Though the end product is a well-proportioned whole, the Chicora Wood house apparently grew in fits and starts. Perhaps as early as 1809, Robert Francis Withers Allston had constructed a small Georgian house, which would sometime around 1830 become the rear wing of an addition that dwarfs it. Even this addition probably came in at least two stages. The distinctive surrounding piazza was completed in the following decade.

The Indian word Chicora may be a Spanish explorer's pronunciation of Cherokee (below).
Oak leaf and acorn adorn this striking plaster medallion (bottom).
A trompe l'oeil hallway is painted on the far wall.
A Duncan Phyfe-style settee is to the right (right).

The house as we see it today is a large two-story gable roofed building on raised brick basement and surrounded on three sides with a piazza. It's been called the "finest" in this region.

Unfortunately, much of the interior mahogany trim and beautifully painted paneling was stripped out by vandals just after the Civil War. But upstairs the plaster cornices and marbleized baseboards remain.

Chicora Wood has been especially fortunate, for it was rescued in the last century by "the woman rice planter" Elizabeth Allston Pringle.

She wrote of her family's home: "It was a wreck—the front steps gone, not a door or shutter left, and not a sash." But being a determined woman, she and her mother spent the remainder of their lives repairing both house and land.

In recent years, the current owners have taken pride in refurbishing their historic treasure. The Historic Charleston Foundation has a conservation easement on the house, thus ensuring that Chicora Wood will never be developed.

The house is on the National Register.

previous spread / A *bright yellow parlor corner furnished in part with a circa 1760 Queen Anne*

oval-topped table, a Dutch inlaid desk with Austrian Rococo mirror above, an elegant wing back

chair upholstered in complementary color, and in the farthest corner, a small case or

"grandmother" clock. But of equal attraction is the surrounding porch that waits just beyond

that opened jib door (page 144, top). **T**he *Chippendale-style chairs with Chinese-lacquer finish*

and round table are reproductions. The painting is of the Withers sisters of nearby Friendfield

Plantation painted by Thomas Sully. The handsome interior colors were chosen by the owners.

While not original to the house, they are authentic to the period (page 144, bottom).

Louis-style armchairs focus our attention on a newly made neoclassical mantel. Painted in 1836

by the prize-winning French artist R. D'Estaing Parade, the subject of the portrait is unknown.

Through the tremendous mahogany doors, we catch a glimpse of the library, or as

Patty Allston Deas fondly remembered, "the book room" (page 145).

this spread / A *pair of neoclassical rockers are placed beside a Victorian worktable. At the window is*

a drop-leaf Pembroke table. The finely carved four-poster bed is trimmed above with hand-painted

material to complement the reproduced French wallpaper and matching curtains.

At the bed's foot waits a recamier or as the laymen have it, "a fainting couch" (left). **T**he *dishes in this*

summer kitchen were found here, as was the plantation-made ironwork. The cook-top stove

and pine cupboard also date from the Allston occupancy. Note the arched window and ash scrubbed

floor. That small, square black iron door fronts a "smoker" (above, left). **I**n *Elizabeth Pringle's time,*

this cabin was her chauffeur's home and together they'd venture off in a "tin lizzie."

And now it's a quite comfortable guest house (above, right).

previous spread / A *garden crossroad and its keepers* (page 148, top). **T***he broadest of porches; the deepest of rivers* (page 148, bottom). **B***oxwood clipped to a playful roundness* (page 149, top). **T***wo symbols of welcome: pineapple urns and Lady Banksia roses on a white picket fence* (page 149, bottom). **this spread / H***all in the old wing: This oldest portion of the house was extensively rearranged. The graceful curves of Windsor chairs seem to be followed by the plaster-hidden spiral of a narrow stair* (left). **N***oted landscaper Lotrell Briggs designed this garden. That rear wing is actually an earlier Georgian cottage* (above).

Using the pen name of Patience Pennington, Chicora Wood's Enderbeth Allston Pringle wrote a concise explanation of her postbellum rice planting. "There are many curious things about the planting of rice. One can plant from the 15th of March to the 15th of April, then again from the 1st to the 10th of May, and last for ten days in June. Rice planted between these season falls a prey to migrating birds—Maybirds in the spring and ricebirds in August and September. It was impossible to plant in April this year, and now every one is pushing desperately to get what they can in May.

Yesterday I went down to give out the seed rice to be clayed for planting today. I keep the key to the seed-rice loft, through Marcus has all others. I took one hand up into the upper barn while Marcus stayed below, having two barrels half filled with clay and then filled with water and well stirred until it is about the consistency of molasses. In the loft my man measured out thirty-five bushels of rice, turning the tub into a spout leading to the barn below, where young men brought the clay water in piggins from the barrel and poured it over the rice, while young girls, with bare feet and skirts well tied up, danced and shuffled the rice about with their feet until the whole mass was thoroughly clayed, singing, joking, and displaying their graceful activity to the best advantage. It is a pretty sight. When it is completely covered with clay, the rice is shoveled into a pyramid and left to soak until the next morning, when it is measured out into sacks, one and one-fourth bushels to each half acre. Two pairs of the stoutest oxen on the plantation are harnessed to the rice drills, and they lumber along slowly but surely, and by twelve o'clock the field of fourteen acres is nearly planted.

It is literally casting one's bread on the waters, for as soon as seed is in the ground the trunk door is lifted and the water creeps slowly up and up until it is about three inches deep on the land. That is why the claying is necessary; it makes the grain adhere to the earth, otherwise it would float. Sometimes generally from prolonged west winds, the river is low, and water enough to cover the rice cannot be brought in on one tide, and then the blackbirds just settle on the field, diminishing the yield by half. I went down into the Marsh field, where five ploughs are running, preparing for the June planting. It is a 26 acre field, very level and pretty, and I am delighted with the work; it is beautiful. When I told one of the hands how pleased I was with the work, he said: 'Miss, de land plough so sweet, we have for do 'um good.' I went all through with much pleasure, though I sank into the moist, dark brown soil too deep for comfort, and found it very fatiguing to jump the quarter drains, small ditches at a distance of 200 feet apart, and, worse, to walk the very narrow plank over the 10 foot ditch which runs all around the field and is very deep.

The evening is beautiful; the sun, just sinking in a hazy, mellow light, is a fiery dark red, the air is fresh from the sea, only three miles to the east, the rice field banks are gay with flowers, white and blue violets, blackberry blossoms, wisteria, and the lovely blue jessamine, which is as sweet as an orange blossom." ✵

From Chicora Wood, a jib door. The best of door and window are combined. In the distance flows the Peedee River (right).

WICKLOW HALL PLANTATION / House Built circa 1835, North Santee River

Built in 1835 by Rawlins Lowndes, Wicklow Hall boasts rarely seen Egyptian revival woodwork. As first constructed, the house consisted of four rooms downstairs and two above, a modest arrangement that lasted until 1912 when the Kinloch Gun Club needed a a temporary clubhouse and added the wing. Cordes Lucas, a descendant of rice mill inventor Johnathan Lucas, was the next occupant. He and his wife modernized the home, a process continued by the Paris family who came next.

Known in the past by several names— The Strip, Duck Point Dwelling, Lowndesville, Hide Out—Wicklow Hall is still a one-of-a-kind retreat (below)."

Distinctive Egyptian Revival trim was used on plantations on the North Santee River. Beyond are twin parlors, each with its own exterior entrance. To the left, the long case or grandfather's clock is Kentucky made. To the right, the table is leather surfaced and was used for gaming (right).

As a young child, the present owner hunted with her father at neighboring Kinloch Gun Club. When her family's ties to that place were severed, she came here to Wicklow Hall. Well known as an artist, conservationist, and preservationist, she remains a hunter. As she explains, "You can't avoid a commitment to preservation if you're truly interested in history. And, after all, most shooters are conservationists." The house is on the National Register. 🪶

This land side entry was a 20th-century addition. The ageless Oriental vases are recently made. The azalea blossoms they hold are only one morning old (left). The Lynn hunt plates of the corner cupboard celebrate hunting, as do the pewter ducks parading on the Centennial mahogany table. The owner has been hunting in this neighborhood since she was 6 years old (above, left). The corner cupboard with rice sheaf medallions is original to the house. The wing back chair is distinctively wide. Coupled with the Alice Huger Smith print above and the great live oak beyond, they suggest a distinctive comfort (above, right).

previous spread / "Not so long ago," the owner recalls, "a bear was ensconced in one of these trees eating acorns as a reclining Cleopatra might have dined on grapes (page 158, top). **T**he owner started life as a painter but is equally adept in other mediums. She did this bronze sculpture of "A Boy With Cat" and has a similar piece in the Brookgreen Gardens collection (page 158, bottom). **P**ineapple finials extend a welcome (page 159, top). **A**ncient brick separates the gardens (page 159, bottom).

this spread / A question of sovereignty. For several years, the mother of this slightly indignant owl has been roosting right beside the house (above). **M**eant to be sent out as Christmas cards, the handsome prints that circle this library are the work of noted wildlife artist Richard E. Bishop. Bishop also did the watercolor of the mallards in a rice field—a scene he'd witnessed while hunting with the owner's father. The two chairs and a game create a quiet corner within an already quiet corner room (right).

WEYMOUTH PLANTATION / House circa 19th Century, Pee Dee River

Like many places in this neighborhood of partly destroyed records, Weymouth's ancient and complex history must be traced through a great length of wills and lawsuits. By 1784, the rice plantation had been increased to almost 2,000 acres and came into the hands of the Izard family. These were a wealthy, highly educated, and well-traveled people, and several of the male members had served with distinction in the recently ended Revolution.

In 1836, a young Ralph Izard, Jr. bought out his sister's share of Weymouth. After finishing his education in Europe and touring extensively, he returned to winter there and to spend his summers in Rhode Island.

He served with distinction as president of the local Planter's Club. His son, who was also well educated, served at Fort Sumter in Charleston, and at the end of the Civil War, he returned to planting. He died in Georgetown in 1891 and the place passed from that family's hands.

Unfortunately, the Izards' Weymouth residence had burned at some point, and so after acquiring the land, the present owners went in search of an antebellum house that they could move onto the site. When none could be found, they turned to a well-known restoration contractor and looked a bit closer to home. A large outbuilding was on the edge of the rice fields. Nearby was an overseer's house that could be moved close to the outbuilding. They built a third structure—a single long room that joined the two buildings together.

The result is an amazingly handsome combination. Heart pine timbers from the first house were remilled into flooring and beams for the connecting room, and the black cypress used both inside and out was collected by the owner in a three-year period before construction.

I must admit I'm particularly fond of this house. The contractor, Tom Graham, is an old friend of mine, and most of the interior woodwork was done by my nephew, Chris Bates, assisted by my son, Aaron. ⚜

These graceful Oriental storks have taken up a permanent residence (below).
At the entry, a little formal garden filled with holly fern, boxwood, gardenias, and an angel (right).

Lady Banksia roses. Beyond, the clever combining of two small houses (left). Spring flowers and miles of rice fields touched by morning light (top). A new steeper roof and cupola for this outbuilding gave a structure to the whole. Sago palms define an opening (above).

In joining together the two existing structures, it was important that one dominated.

Hence, the shallow gable roof of the outbuilding was altered to a steep pitched hip and a cupola was

placed atop that. This intricate cypress stairway was hung beneath the cupola, which created a

sun-filled high ceiling centerpiece (left). The beams and flooring in this sunroom were milled from the

roof structure of the altered outbuilding. The sashes of the tall windows were salvaged

from another plantation (top, left). Seven kinds of wood went into this Windsor chair.

The local craftsman who made it went into the forest, selected the trees, and hand split the material.

Run in an uncommon horizontal, the beaded pine paneling is original to the building (top, right).

The owners collected cypress for three years, hence these panels. The rice bed

is a skillfully carved reproduction (above, left). The warm glow of cypress meets the warm glow of

heart pine in this kitchen. The sweet-grass basket is a local art form. The owner took lessons

from a gifted blacksmith and built that pot rack (above, right).

ATALAYA / House circa 1928, The Atlantic Ocean

Philanthropist Archer Huntington based his design for a winter home on a Moorish Castle he'd seen in Spain. And he kept that design in his head, working without plans and announcing his intentions as they went. "Mr. Huntington," said contractor William Thomson, "if you tell me much more, I'll find out what you're building." A laughing complaint, for in those days of the Great Depression the

Genevieve "Sister" Peterkin recalls visiting as a child when Indian blankets hung from the walls and the furniture was austere in this breakfast room and master bedroom (below). Because of threats to Huntington's wife, wrought-iron guards were placed on the windows (right).

local labor force was happy to have work and dozens were trained on site as masons. The dripping mortar style, an excellent one for beginners, was called "the Huntington Squeeze."

The resulting building measures 200 by 200 feet and consists of relatively small rooms surrounding an immense central courtyard. A flat roof, walls of solid fireproof masonry, brick floors, and iron barred windows—this Moorish design makes for a curious Carolina beach house. Indeed, now empty of furnishings, it's difficult to imagine the house as being comfortable, and yet it is remembered so.

Anna Huntington sculpted the aging horse for Brookgreen's "Don Quixote" in the attached studio.

"During the 1930s, dozens of wealthy Northerners moved into the Lowcountry and built or rebuilt antebellum mansions in an attempt to recapture the glory of the Old South. Only one couple chose to emulate the glory of Spain in the days of El Cid. It's an architectural folly, but only in the kindest sense of those words—a building so outrageously romantic that it defies all logical complaint." I wrote that in 1987 and it's still true today ❧

previous spread / Lowcountry palmettos in the courtyard take the place of Moorish palms.

Actually, palmettos in the wild don't grow this far north (pages 170-171).

this spread / A Moorish fortress nestled among the dunes of a Lowcountry beach (top).

A place of romance, Atalaya is now open to the public. Atalaya translates

as "watchtower" (above, left). **T**his straight axis road once ran from the center of

Brookgreen Gardens to the center of the Atalaya impoundment (above, right). **T**he Quixotic home

of a generous couple (right). / Photograph by Alice Turner Michalak

BROOKGREEN GARDENS / ESTABLISHED CIRCA 1936, WACCAMAW RIVER

With the coming of the Great Depression, Southern agriculture slid even further into ruin. A duck hunting club was set up here and failed. Then Archer Huntington spied the newspaper advertisement for 6,300 acres of South Carolina rice plantation. The son of a railroad magnate, Archer had recently married Anna Hyatt, a skilled and, for the most part, self-taught sculptress. She was suffering from tuberculosis and they chose Brookgreen as a place to rest. Once here, the couple was so impressed by the beauty of the abandoned house site and surrounding woods that they reached a momentous decision. As memoirist Genevieve "Sister" Peterkin recalls, "The joke that Archer Huntington told on himself was every where he put his big foot down a museum would spring up, which was almost true, and actually that is what I remember most about him—a very large, well-dressed courtly man with very big feet, size fourteen at least, but then I was a very little girl when he arrived. ... The work was in the classical tradition—realistic human figures and animals." The sculpture, the live oaks, the flowers—Brookgreen Gardens, she points out, has justifiably been mistaken for heaven. "Its beauty can be overwhelming."

Now heading into its fourth decade, Brookgreen Gardens wishes more than ever to offer its visitors a broad and enjoyable aesthetic and educational experience. With this in mind, a new Lowcountry center is being built, one that will dwell in part with the African contribution to rice cultivation and offer a better understanding of our natural surroundings.

Boat and carriage rides have been added, the garden planting has taken a more year-round emphasis, and the sculpture exhibits have broadened in scope, with many well-received traveling shows now offered.

Most recently, tours of a Southern Living residential cottage have given tens of thousands of visitors insight into a new, more livable Lowcountry architecture.

The "Great Chain of Being" gracefully incarnated (below). / *Photograph by Alice Turner Michalak*

A *bend in the garden path* (right).

previous spread / As curator Robin Salmon explains, "Huntington designed this 'Diana'

as a series of curves that would draw the eye up from the base to the already shot arrow." The piece

was one of her most popular (page 176, top). **S**wedish born artist Carl Millis believed that sculpture

should be enjoyed. He exchanged the "stuffy" approach for an aesthetic that would make people

smile. These figures represent the practitioners of the arts leaping across the waves with little flying fish

spurting around their feet (page 176, bottom). **L**illies and more lilies (page 177, top).

Young Samson rends the lion. Note the length of his hair (page 177, bottom).

this spread / A new addition, this "Diana" suggests a more delicately balanced approach

to maidenhood (above). **W**ith an increasing emphasis on education, a children's corner invites

the next generation of garden enthusiasts (right).

HOBCAW BARONY, BELLE BARUCH FOUNDATION INSTITUTE / WINYAH BAY

The first European colony on the North American continent is thought to have been here on the Waccamaw Neck. In 1526, the Spaniard Lucas Ayllon watched over the decimation of his 500 followers until he died. Mutiny followed, and the slaves he had brought revolted. In a homemade ship, 150 slaves managed to return to Cuba. In 1718, the site was granted in a 14,000-acre block to one of the Lord Proprietors and was then divided into rice plantations just before the Revolution.

Some of these aging buildings date to slavery times. But the little chapel was built by Baruch for his workers in the 1920s (below).
The most avid of duck hunters, Mr. Baruch built this sturdy home beside a bay that teemed with waterfowl. The first house burned in a dramatic Christmas Day fire, and this 1930 fireproof replacement is brick (right).

In 1905, entrepreneur and statesman Bernard Baruch began to reunite the parcels and used his reformed barony as a place to hunt, relax, and entertain friends.

Baruch's daughter, Belle, inherited the land and willed the property to the state for use as a teaching and research center.

Today, its forestry and marine resources projects are operated under the joint control of Clemson University and The University of South Carolina.

SOUTH ISLAND—YAWKEY CENTER / ESTABLISHED 1976, NORTH SANTEE RIVER

Learning at the age of 16 that he had inherited a great fortune, Tom Yawkey replied, "I hope I'll be able to do some good with it; I hope I'll be as good a man as my dad." At his death three decades later, all agreed Mr. Yawkey had done just that, for the years between had been spent in generous and charitable works of a broad nature.

The wildlife preserves of North and South Island are one such endeavor. Located to each side of the Georgetown harbor entry, this property was a part of the inheritance from his father (who was actually his Uncle Bill). An avid hunter, Tom Yawkey spent the remainder of his long life overseeing their management for a variety of wildlife. At his death, he was determined that the vast acreage would become a refuge. With a $10 million endowment, the Yawkey Center has in the years since continued to protect wildlife and educate others in the field of conservation.

Though North and South Island had been used for the usual sea island enterprises of cattle raising and ship stores manufacture, rice became the major crop.

On the south side of the harbor, miles of dikes and canals were eventually built. And as was the usual case by the 1890s, Northern duck hunters were finding their way to the area. At this point,

This handsome house beneath the oaks is lived in by the manager and his family (below).
A *luminescent egret rises from the shadows* (right).

Gen. Alexander Porter bought the properties and began to promote them. President Grover Cleveland came to hunt frequently and others followed.

In 1911, Porter sold the properties to Will H. Yawkey, Tom Yawkey's guardian and uncle. As a boy, Tom had come to hunt. He kept returning from then on. Upon his death, this tremendous wildlife preserve was established.

Though the migratory waterfowl population is still the center's chief resource, it should be noted that a haven has been provided for much, much more. A variety of hawks, as well as osprey, peregrine falcon, and golden and bald eagles also attend. The bald eagles, in

previous spread / M*orning flight* (page 184, top). **T***he brackish water supplies the numerous alligators with a steady diet of crabs, fish, and shrimp* (page 184, bottom).

T*wo wild turkeys slipping away. This food patch on Belly Ache Road is maintained for their convenience* (page 185, top). */ Photograph by Alice Turner Michalak.* **M***ore than 44,000 acres of rice fields were once constructed within this area's river systems. And here they're still being maintained for the benefit of migratory waterfowl* (page 185, bottom).

this spread / W*hite ibis enjoying a rice field* (left). **T***his rice chimney is a brick relic of the Lowcountry's "industrial age"* (top, left). **A** *blur of wings* (top, right). **S***unrise* (above, left). **K***ingfisher and rice field trunks. Such antebellum water-control structures probably evolved from hollow logs or "trunks." A rising tide is allowed to enter, a falling tide is allowed to leave* (above, right).

following spread / B*eams salvaged from a rice barn form the ceiling of the trophy room. The 1930s furniture was designed by Molesworth and manufactured in Cody, Wyoming* (page 188). **T***om Yawkey owned the Boston Red Sox. Here, a finely penciled Ted Williams waits his turn at bat* (page 189).

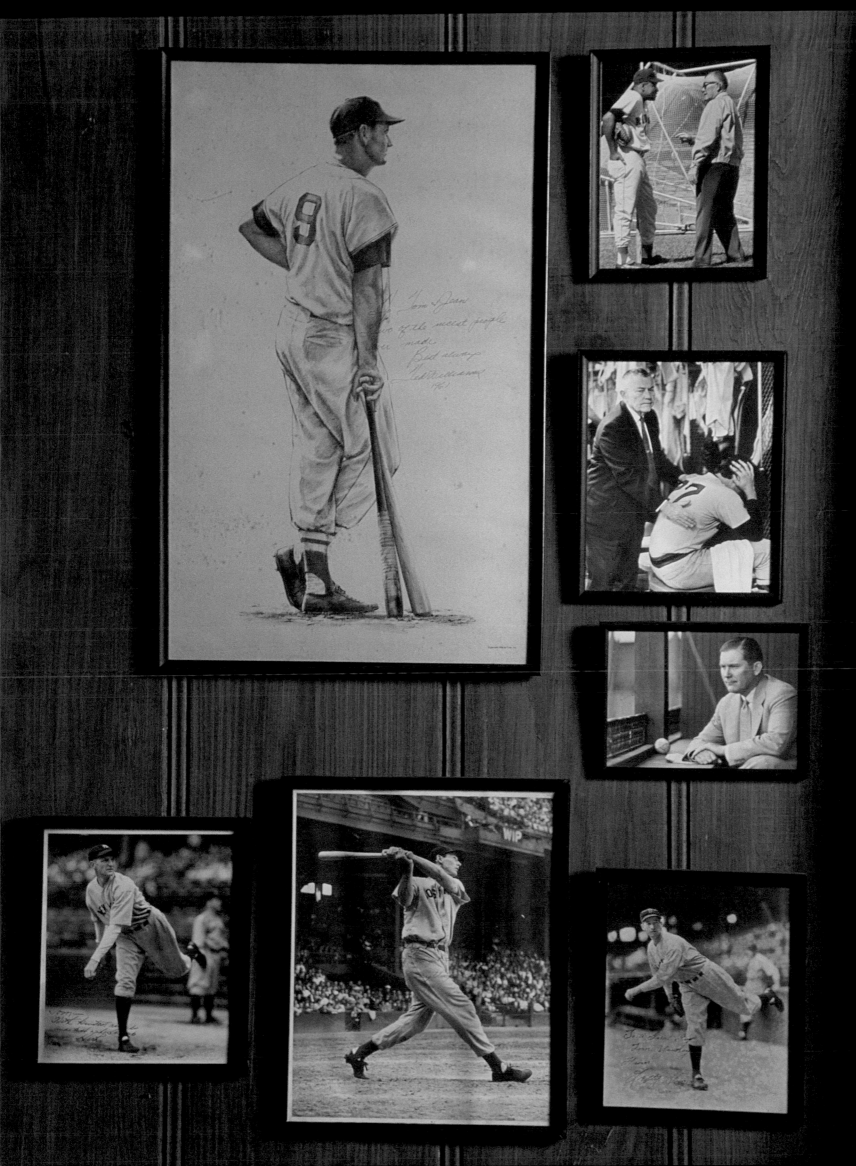

particular, are a common sight with eight active nesting pairs. The red-cockaded woodpecker, an endangered species elsewhere, has eight strong nesting areas established. In addition, fox squirrels, otters, alligators, and loggerhead turtles are protected, with the endangered turtles averaging 250 nests a summer.

As manager Bob Joyner explains, this vast protected acreage is ideal for the conducting of field experiments. Chemical contamination (especially from insecticides) has been kept to a minimum, there are few people about, and a substantial amount of already accumulated data exists for scientists to draw on. The various research projects tend to complement each other, and the work being done just to the north at Hobcaw dovetails in, as well.

Indeed, what makes the Yawkey refuge an especially effective ecological haven is its location in the midst of other such areas—the Santee Coastal Reserve, the Cape Romain National Wildlife

———

This woodland's chapel is still used on rare occasions (below). *In 1928, this small AME church was built for the island's black families. As member Martha Seargent fondly recalls, one wedding reception was so big, they had to send back to the mainland twice for store-bought ice cream* (right).

Refuge, Capers Island, the Francis Marion National Forest, and the Baruch Institute at Hobcaw Barony.

In recent years, biologists have begun to understand just how important such a massing of acreage is. Studies have shown that small holdings, though still valuable, tend to "erode" at the edges. That is to say, they do not maintain their protective efficiency and are abandoned by the very species that they hope to protect.

Obviously, then, the Yawkey Center must continue to be a refuge for protecting and for experimentation so we may continue toward a realistic understanding of how animals and plants behave—to learn how ecology actually works.

For many years, my father, William P. Baldwin, Jr., was involved with this institution, first as a wildlife management consultant and then after Mr. Yawkey's death as a member of the center's board of directors. ❦

Between the Indian wars, the American Revolution and the
Civil War, it would seem that all of the Lowcountry has been a
battlefield. Still, the Beaufort area suffered most—especially in 1865
when Sherman's troops entered the state from Savannah and burned
their way northward. From the ashes does come a redemption of sorts.
Handsome postbellum plantation houses have been erected, and the
rivers and rice fields of this particularly isolated section are
receiving the best of environmental protection.

Beaufort and Her Environs

The ACE of the ACE Basin Project is an acronym for the three rivers making up this watershed—the Ashapoo, the Combahee, and the (South) Edisto. Approximately 350,000 acres of forests, wetlands—both fresh and brackish—and saltwater marshes, plus barrier islands and beaches are being conserved through the cooperative efforts of private landowners and various public and nonprofit organizations. Put as simply as possible, to initiate a conservation easement, a private land owner donates to a nonprofit organization like The Nature Conservancy the development rights to his property. An appraisal on the property without restrictions is made. Then an appraisal with restrictions is made. The charitable gift is the difference between the two appraisals. The owner may then deduct that gift from his federal and state taxes. Estate taxes can also be reduced.

For instance, if you have 1,000 acres, you limit yourself to three houses. Only three may ever be built on the entire property. The land was worth $3 million, and now it's valued at $1 million. You get the difference as a write-off. And the agency has responsibility for enforcing the terms of the agreement. Those terms can vary. Not just the subdivision of property can be considered. At the owners wish, a timber management plan may be involved and the maintenance of fields and dikes. Or they may wish for the prohibiting of hunting or the preservation of wetlands—this last, of course, being subject to state and federal laws.

Two obvious questions "drive" an easement. What are the conservation needs of that particular property? For instance, does the conservation agency hope to preserve natural plant communities? Or is the wish to provide habitat for migratory birds? And, of course, equally important, what are the needs of the owner? What use does he wish to make of his land? An easement is tailored individually in answer to these questions.

With the success of the ACE Basin, similar conservation efforts are being pursued elsewhere in the Lowcountry. Winyah Bay at Georgetown and the Sewee to Santee region to the north of Charleston are two such focus areas. Like the ACE Basin, these major coastal watersheds were at first valued as waterfowl habitat, but a more complete ecosystem approach is evolving—from ducks and geese, the game species, to all migratory birds. The natural plant communities like the long leaf forest or maritime forest and fisheries resources are considered. The wetlands are particularly valued.

At this point, the U.S. Fish and Wildlife Service, the state wildlife organizations, and nonprofit groups like The Nature Conservancy and Ducks Unlimited are involved. The major building blocks for all of these projects are the old plantations

In that sense, the South Carolina Lowcountry is unique. This is the only place on the Atlantic seaboard where rice was planted, and many of the large parcels are still intact. And realistically, these plantations will eventually go one of two ways: They will be conserved or they will be developed—cut into small parcels and built upon.

True, these plantations aren't pristine landscapes. They're still being managed in a traditional fashion. But the basic ecological processes remain intact. The land is there, and a philosophy of sport hunting is now broadening into a conservation ethic and with cooperative partnerships and easements like those used in the ACE Basin, there's every reason to be optimistic.

In terms of conservation, the future for plantations is bright.

Viewed from Mackay Point Plantation. The Combahee marshes—a glory of the Carolina coast.

WILLIAM SEABROOK PLANTATION / House circa 1810, Steamboat Creek, Edisto Island

The Edisto appendix of Ramsey's 1808 history reports: "A census of the Island ... would rate the white population at 236 inhabitants. Of these 111 are males, and 135 females. Of the males 37 are married, 4 are widowers, 9 native of Europe, and 2 of the middle States; of the females 37 are married, 12 are widows, and all are either natives of the Island or the adjacent parts of the State. ... The island consists of almost 30,000 acres which much of this being cultivated for sea island cotton. The masters treat their large slave population with notable kindness. All the white inhabitants are remarks on as being industrious, friendly and fair and liberally support their religious institutions. Dancing is not encouraged as it is elsewhere— they tending to be a large Presbyterian element. Boat sailing and fishing take the place the horse racing and hunting of the mainland."

Only the alarming death rate due to fevers kept this place from being a perfect retreat from the world, and it was thought that the draining of swamps and the movement of "summer houses" to new locales would cure even that.

Self-contained and romantically self-involved, this was the island world of which William Seabrook was a very big part.

At 17, Seabrook took over both his and his mother's estate. Experimenting with salt marsh mud as a fertilizer, he was one of the very first people to gain great wealth from the raising of sea island cotton. In addition, he established the steamboat line that connected the island to Charleston. He also owned several other large nearby plantations.

A pair of garden nymphs guard this entry (below). *An all-enduring conservation easement guards these grounds* (bottom). *Planter William Seabrook is handsomely memorialized in brick and timber. Note the initials* (right).

this spread / **D**istinctive are these broad elliptical arches—both inside and out (left).
A serpentine limb of a live oak tree (above).

following spread / **A** Union soldier embellished this bedroom wall with the mighty
power of the Union Eagle and artillery (page 200, top). **M**ore graffiti from uninvited
guests: "Harry Crouse, Pittsburg, Penna. Armys ISS Regt. P.U., Edisto Island, June 26,
1862 (page 200, bottom)." **T**he portrait is of Elizabeth Dodge, Donald's great,
great grandmother. The camel-back sofa is neoclassical in style. The needlepoint
on the fireside bench presents Odo and Gilly (a dog and game manager)
hunting at Pon Pon (page 201).

In 1810 Seabrook began construction of "his mansion house." A persistent rumor names Hoban as the architect (the designer of the White House in Washington, D.C.), which is possible but certainly not proved. Still, some expert hand was guiding the building project. The floor plan is the traditional four rooms over four with central hallways, but in treatment all is exceptional. The two-tiered, Federal-style portico presents a set of particularly graceful arches—enriched further by broad windows and ornate detailing. The interior is treated in a similar vein. Following its construction, the house's design influenced many others.

Graceful curves: a conspiracy of swags, fan, banister, and wallpaper (below). This faux grained door was designated in Union Army pencil to be the entrance to "Room No. 1" (bottom). Portrait of Gertrude H. Dodge over the sideboard. Sheraton side chairs are pulled to a Regency-style pedestal table (right).

The property remained with the Seabrook family after William's death. Federal troops used the house for staff headquarters and left behind graffiti to commemorate their stay. At the war's end, a new set of calamities began, and the Seabrooks parted with the home.

Gertrude and Donald D. Dodge bought the William Seabrook Plantation in 1929. During the 1930s, they restored it to its earlier splendor. After the deaths of the Dodges in 1974, their four children and their families continued to enjoy the Seabrook house. In 2000, the family sold the plantation to a Charleston family with provisions.

Rather than seeing the acreage broken up into more residential developments, both parties agreed to make easement conveyances a part of the sale. The Preservation Society of Charleston helped with the stewardship responsibilities of the house, and the Lowcountry Open Land Trust did the same for the acreage. Both easements are in perpetuity and will assure that the plantation will be preserved in its current historically and environmentally important condition.

The house is on the National Register. 🔹

DAVANT PLANTATION / HOUSE CIRCA 1820, GILLISONVILLE

For services rendered to King George III, in 1770 John Hobard was granted 100 acres of forested land. He gradually cleared and farmed this land. In 1811, his granddaughter sold the property to Stephen McDonald, who sold it in 1813 to William Watters. In 1815, Watters sold the property to John Cheney, who christened it "The Oaks." Cheney's daughter married R.J. Davant, who bought the property and changed its name to Davant.

This mansion was constructed around a farmhouse that was overlooked by Sherman's troops (below). *Those oaks, seen from the dining room window, are centuries old* (bottom). *"A silly old tin angel," says the gardener. The rugged chair is plantation made.* (right).

Four hundred acres were added, and a handsome house, kitchen smoke house, stables, and laundry house were built. Besides enjoying success as a planter, Davant was a practicing lawyer, a public commissioner, and a vocal advocate of Secession. He was serving in the South Carolina Senate when Sherman entered the state with his troops. His house was burned, and as family lore has it, dead hogs were stuffed into the well. But a smaller residence located "across a boggy cut" and some 2 miles away was missed. Davant had this dismantled piece by piece and rebuilt on this site. It forms the basis of the present house.

From Davant's descendants, the house passed to the Berolsheimer family—founders of the familiar Eagle Pencil company. They trained English pointers and Irish setters that competed successfully on the field trail circuit. In addition, The Berolsheimers were avid quail hunters. They increased their holdings to 2,500 acres and leased an additional 500.

To improve the existing house, they hired well-known Augusta architect Willis Irvin. With his guidance, they added the bedroom

previous spread / To the left is one of Alice Ravenel Huger Smith's drawings for The Dwelling Houses of Charleston. *Cotton bolls are showcased in a Japanese lantern (page 206, top).* **T**he owners have been collecting Southern art for many years. The result is a museum-quality exhibition in their home. Marty Whaley Adam's "Evening" hangs in the far corner. "The Bend in Church Street" was spotted in a New York gallery. The quail perched upon that end table is by carver Floyd Robbins. The Remington "Bronco Buster" is a Western interloper (page 206, bottom). **T**he previous owner, Myra Berol, designed and painted the wall covering herself. It forms the backdrop for a family silver service and ormolu candlesticks (page 207). **this spread / T**his laundry building serves as the plantation's office. The "standup" desk was used by an ancestor, former governor of the state R.F.W. Allston, at his office in the Fireproof Building in Charleston (left). **N**ote how the portico's flattened siding allows the intricacy of the arched embellishment to shine. Architect Willis Irvin was at home in this neoclassical tradition (above).

wings and kitchen and replaced the beaded siding of the interior with plaster. In addition, the existing outbuildings were refurbished, and five new tenant houses and a new stable and dog kennel were added.

The present owners have carried on the tradition of maintaining and improving the property. For them, however, Davant is operated as a "working plantation." They harvest timber, and most important, they offer recreational hunting. In order to have an abundance of deer, dove, quail, and turkey, fields and numerous feed plots on the 2,500 acres are planted with corn, millet, sunflowers, partridge pea, chufa, clover, and sorghum.

*previous spread / **A** spring border of poppies, pansies, and verbena* (page 210).
***I**n an earlier life, the garden house was storage for railway trunks* (page 211, top).
***T**he owners are collectors of Southern art, and their tastes are happily eclectic* (page 211, bottom).
this spread / S*cene from the back garden* (below).
***T**his guest house was converted from the old kitchen* (right).

The plantation served as the location for the movie *Something to Talk About*, which starred Julia Roberts, Dennis Quaid and Robert Duvall and was about a modern-day family of plantation-based horse breeders. The owners found that experience to be altogether enjoyable. And they still recall with a smile Robert Duvall's alarmingly accurate and humorous portrayal of a Lowcountry plantation's patriarch. ⚜

COMBAHEE PLANTATION / HOUSE CIRCA 1872, COMBAHEE RIVER

Known originally as Hamburg, this property was held by the Heyward family in the mid-1700s. At his death, Daniel Heyward (owner of Charleston's Heyward-Washington house) left it to his third son, James.

In 1794, James married Susan Cole. Though said to be the daughter of a Welsch butcher, Susan

This narrow lane beneath the avenue suggests the informal quiet of what's to come (below).

***S**unrise over the Combahee (bottom).*

***T**hese projectile points, drills, and pottery were discovered on these grounds and are displayed in a 19th-century curio box table (right).*

had a sister married to the Earl of Berkeley, and her maiden name might actually have been Tudor.

As historian Susan Linder explains, "She was said to be beautiful, witty, charming, and to have talent as a writer of dramatic works and as an amateur actress. Family tradition says that upon his return to South Carolina, James' wife 'was received by his family to say the least ungraciously,' and that he soon died of a broken heart, leaving his property to his wife for her lifetime, after which it would revert to his brother, Nathaniel."

Then, according to a published family history, Nathaniel investigated his brother's marriage and alleged that, between 1784 and 1794, Susan had been the mistress of five different men.

The will was not overturned. After a year of widowhood, Susan married Englishman Charles Baring. He was 10 years her junior.

Baring challenged Nathaniel to a duel. They exchanged shots at a point halfway between their homes, and that matter of honor was considered settled.

Susan lived almost a half-century longer and helped to establish the mountain retreat of Flat

This dining room is furnished with a variety of handsome Chippendale- and Hepplewhite-style pieces.
But of particular interest are the playful juxtaposing of more modern art. The two leopards
above the mantel are by Henri Maik, and "Three Apples" is by Rutling (left).
The owner's design philosophy is a rather simple one—"Just comfortable.
I don't want to worry about grandchildren running around." This well-lived-in room
features a banjo barometer and deer head, a country-style ladder-back chair, a
Chippendale-style card table, a child's Windsor chair, and a pair of Queen Anne side chairs.
Beyond all that is a particularly sunny breakfast nook (above).

Rock and, most particularly, aided in the building of Flat Rock's handsome little chapel now called St. John's in the Wilderness.

During those summer retreats to the mountains, filled with gracious receptions, dinners, and balls, Susan was, Linder writes, "remembered for her elaborate dress, a headgear of graceful plumes, flowing ribbons and sashes—usually blue—and magnificent diamonds."

The spiral of wrought iron no longer beckons. The gates been opened (below). Migrating Canada geese were once sport, but are now decorative only (bottom). Random-width boards of pecky cypress rubbed with white milk paint follow the curve of the rising stair. The French tapestry above is a family heirloom (right).

On her death, Baring's relatives attempted to retain Hamburg, but the property reverted to the Heyward family as James' will had specified.

Young James B. Heyward, "a plain man, simple in manner, courteous in deportment," owned the plantation at the time of the War and was remarkable in the fact that in the period of Reconstruction he was able to harvest millions of bushels of rice while his neighbors were going bankrupt.

In 1872, he replaced his Sherman-burned home with the present large frame structure and continued to farm successfully until the Great Hurricanes of the 1890s destroyed much of his fields.

In 1913, the mortgage was foreclosed on by a company attempting to raise upland crops in the area. Felix DuPont, the company's largest stockholder, acquired Hamburg, renamed it Longview, added the drive and the young oak allée, and attached a cottage to the side for a kitchen and extra bedrooms.

The Lane family came next, adding the bay windows and screened porches.

Then the property was sold and partially subdivided. The present owner received the house and extensive grounds, to which they gave the name Combahee. 🪶

TWICKENHAM PLANTATION / House circa 1880s, Combahee

This property appears in the earliest records as belonging to "that worst of all" Indian trader Joseph Bryan, but by 1744 Walter Izard was the owner. And through death and marriage, the two tracts were passed through the generations (the records were burned) until one portion came into the hands of James Reid Pringle in 1820. The Bull family owned the second portion. Beyond that, little is known. Quite possibly both pieces were being managed by overseers and did not serve as plantation homes.

The name Twickenham appears in print for the first time in regard to the marriage of Sarah Heyward to Thomas Middleton Hanckel. Trained as a lawyer, Thomas and his wife continued to live in Charleston, and with some success, he watched over the rice planting from afar.

Sherman's troops burned whatever house did exist. And in 1871, J. Bennett Bissell leased the fields for planting. At the end of eight years, Hanckel lost Twickenham to Major John H. Screven at a sheriff's sale.

Screven built the present house, and on his death in 1903, his grandson Robert Turnbull inherited the property. He remodeled the house in 1929 and made it his home. ✍

*R*enewal. *Sherman's troops burned the original house, but a new start was made using the same foundation* (below). *T*hese *treasured Audubon prints came from an ornithologist grandfather. The decoys were collected by the generation that followed* (right).

*this spread / **C**hippendale-style chairs complement the tulips, and vice versa* (above).
The portrait in this spacious hall is of Kate, a yellow lab kept by the
preceding generation. Now bassets have the floor (right).
*following spread / **O**rder and proportion under a blue sky* (page 224, top, left).
***O**ld Glory* (page 224, top, right). ***U**sing God's ingredients, landscape designer*
Robert Marvin worked his magic (page 224, bottom). ***O**nce the home of beloved*
cook Lily Farr, this "Lily Pad" now serves as a guest house (page 225, top).
***A** ribbon like grace from metal—and from the oak limbs,*
as well (page 225, bottom, right). */ Photograph by Alice Turner Michalak*
***D**ouble the pleasure from these double "hose in hose"*
azalea blooms (page 225, bottom, left). */ Photograph by Alice Turner Michalak*

BONNY HALL PLANTATION / HOUSE CIRCA 1897, COMBAHEE RIVER

The Bonny Hall house was built by George W. Egan in 1897. A late Victorian retelling of a Georgian manor, the floor plan is of the traditional central hall containing the stair with two rooms to each side. The windows, particularly large and numerous, are wood-sided but rest on a high brick foundation. The roof is hipped and covered in slate. The interior was plastered with wainscoting of the familiar Victorian beaded board, and the mantels, too, are suggestive of that era.

Having purchased the land from J. Bennett Bissell, the most successful of the Reconstruction-era planters, Egan continued to grow rice commercially. This venture finally collapsed, and George Lyman, the next owner, made some repairs and began an extensive camellia garden. He used the property for hunting only. On his death, the Doubleday family acquired Bonny Hall for hunting. At the time, Doubleday was the largest publishing house in America and ample funds were available for an extensive makeover.

Noted landscape architect Umberto Innocenti designed a new garden and moved the driveway and entry around to the rear of the house.

Extensive renovations to the house were begun under the direction of noted Charleston architect Samuel Lapham. A large wing was added to each side of the house, the central entry hall was done away with, and much of the Victorian interior was given a Colonial Revival treatment. In addition, several new outbuildings were added, and the old stable was rebuilt.

The present owners, though happy with the additional space of the wings, removed most of the Doubleday "improvements." The house now has its entry hall once more, and the entry door to the rear has been sealed off and the drive returned to where it began. The Colonial Revival mantels and some of the trim werere-

A rose-filled springtime entry (below).
This cupboard over drawers chest is in the newly reclaimed entry hall. Made in New York in the Sheraton style, the piece's mahogany drawers are dressed in book-match crotch mahogany and complemented with radiating panels of a darker burl. The portrait is of a family member. Lapis, the Australian shepherd (named after the color of her striking blue eyes), is sleeping on the entry stoop (right).

this spread / A*bove the mantel is the portrait of a specific Bonny Hall eagle done by artist Robin Hill. And on the mantel a miniature Japanese band is displayed. The inlaid teak box on the table is of Chinese origin, as are the plates on display* (left). **B***uilt in 1897, the home was built as a planter's residence* (top). **D***eji the "near horse" (to the left) and Theo the "off horse" in harness to a Kuhnle Marathon vehicle. The owner competes in combined driving—an exciting horse-drawn equivalent to sport car rallying* (above).

following spread / T*he Doubleday family added this rustic kitchen wing, which the present owners brightened considerably. The corner cupboard is a New York piece, the oak table is English, and the chandelier is Dutch* (page 230, top). **I***n the master bedroom, the owners removed a wall, converting two rooms into one, and added the large window that looks out on the camellia garden* (page 230, bottom). **T***he chairs, table, and chest in this dining room are English, and the oak cupboard is Welsh made. The silverware is Irish. The porcelain is Chinese, a cabbage leaf motif designed for export. The mirror is from the Regency period and probably American* (page 231).

placed with something more appropriate. Most important of all, perhaps, was the advice given by famed landscaper and architect Robert Marvin. "Let a little light in," he said. At the kitchen wing this was simply a matter of trimming the shrubs and trees. But on the bedroom wing, a large expanse of windows was added. And throughout the house, curtains were kept to a minimum. The result, now decorated with many family heirlooms, makes a comfortable home for both the owners and their great assortment of pedigree dogs.

Famed New York landscape architect Umberto Innocenti laid out this garden in the '20s. In more recent times, the Lowcountry's Robert Marvin aided in its restoration (below).
A terra-cotta maid of ancient Italian origin (bottom).
This entry door has been converted to a broad window and accompanying window seat (right).

Bonny Hall has enjoyed some rather noteworthy residents in the past. "Sold to Hugh Bryan" is written on the back of one of the earliest deeds, and apparently it was here or close by that Bryan in 1738 began his preaching among the slaves and Indians.

As a boy, Bryan had been captured during the Yemassee uprising and held prisoner. He eventually became a respected planter and churchman, but under the influence of his wife (the daughter of Indian fighter Tuscarora Jack) and the masterful preacher George Whitefield, he began to educate his slaves to the teachings of the Bible. His neighbors, deciding he was inducing slave revolt, had him arrested. He was soon returned to the life of a prosperous planter.

While his actions brought ridicule and hastened the failure of Whitefield's Great Awakening in South Carolina, Bryan continued to educate his own slaves, an action that would eventually lead to the establishment of black churches in neighboring Georgia.

On a related subject: While living in the Doubleday Bonny Hall guest house, noted English author W. Somerset Maugham wrote *The Razor's Edge*—a novel that dealt with his own quest for meaning. ≈

ORANGE GROVE PLANTATION / HOUSE CIRCA 1928, CAPERS CREEK, ST. HELENA ISLAND

In 1753, Peter Perry purchased 473 acres of undeveloped land on the back side of St. Helena Island. Over the next 12 years, he transformed the property into profitable Orange Grove Plantation. Indigo was the crop, one tended by 46 slaves. At Perry's death, the island's largest indigo producer acquired the place and continued to produce the dye and to raise cattle, sheep, and hogs.

Beyond this, the records are sketchy. During the northern occupation, St. Helena Island was divided into 10-acre tracts of land and distributed to the former slaves under a program known as "The Port Royal Experiment." Over the years, some of the island plantations have been reassembled, including this one.

In 1928, Henry Bowles of Massachusetts purchased Orange Grove for a hunting retreat. He had the original house torn down and commissioned a Cape Cod-style home with arched front and back porches and an interior of pecky cypress and painted moldings.

Four decades ago, the present owners began their residency and expanded the house as their family grew. The rear porch was enclosed and incorporated into the living room. The original kitchen was transformed into a glass sitting room, and the east wing was expanded for more bedrooms. Recently, Robert Marvin designed the new garden room and deck to complement his redesigned and enlarged enclosed garden.

The rear of the house is now encased in glass (below).

Following the advice of landscaper Robert Marvin, this sunroom addition to the rear of the house "brought the outside in" and gave a view of the river as well (right).

previous spread / O*range Groove turns this traditional face to the approaching visitor (page 236-237)*

this spread / A *touch lower than most, this "comfortable" table is 18th-century English and*

a family heirloom. Dresden china and Chelsea china are displayed on the shelves of a circa 1770

Chippendale bookcase. A circa 1790 sideboard is on the far wall. The wallpaper is hand done by

Gracey. Well-known artist Charles Fraser painted these local "Orange Grove Ibis" (left).

T*his massive mantel is all that was salvaged from the site's antebellum house.*

Now that the back porch has been incorporated into the room, the mantel is a bit more to scale.

The bird prints are Mark Catesby originals. The rug is from India.

The coffee table is papier-mâché on a faux bamboo stand.

The sunny fabric is by Brunschwig (above).

Darwin's anomaly—a parading peacock (top, left). **S**hadows get tossed by light in these green waves (top, right). **C**anadian geese nesting (above, left). **M**odern farming—tomatoes grow here, where indigo and sea-island cotton were once abundant (above, right). **A** garden room extension of the new kitchen. The chair fabric by Baussae is in honor of the several banana trees growing just outside (and taken in for the winter). The interior decorator suggested the slate color for the walls to make them disappear, which they do in a way (right).

PAUL AND DALTON PLANTATION / HOUSE CIRCA 1928, COMBAHEE RIVER

Going under many names, this large property was held during much of the 18th and 19th centuries by three consecutive women: Rachael, Ann, and Margaret Stock. Each outlived her husband—from 30 to 50 years—and on her own added to the acreage.

In 1859, James Paul bought a portion, but there's no record of him making a crop before the sailors of Union gunboats raided the plantation and burned the house and outbuildings. He died in 1866, and 12 years later, Paul plantation was sold at a sheriff's sale to W. Dalton Warren, who already held adjoining Dalton plantation. (Family legend and a very real tax receipt suggest a $71.98 tax lien cost them the 3,000-acre plantation.) The combined pieces were sold several times before going to the Charles Lanier Lawrance.

Lawrance, businessman and inventive aeronautical engineer, had a passion for hunting. In 1927, he acquired a Lowcountry plantation where he hunted quail. Needing duck marshes, he bought

Designed by the former owner, an aeronautical engineer, this 1929 lodge seems the product of a far earlier time (below). Gary Moff's painting of widgeon set in upon a Paul and Dalton rice field hangs above the living-room mantle. The faux paneling and post and beam of the wall suggest the lodges of a slightly earlier era (right).

Paul and Dalton. In 1928, he designed and built the present hunting lodge—a relatively small building. The middle or main living area is set well up on three great brick arches, and to each side is a lower bedroom wing. Tall chimneys and an even taller cupola with a distinctive weather vane add to the mass of the center.

According to historian Suzanne Linder, my father, William P. Baldwin, Jr., acting as a broker, advertised the plantation as an operating business in 1961. The rice fields were kept drained and used as pasture, with alligator weed providing more than enough feed for the 100 head of cattle. And he wrote, "This property provides very excellent duck, deer, turkey and quail hunting."

Today's owners have added a guest quarters of a fittingly Lowcountry design and use Paul and Dalton as a hunting retreat. ⚓

With exposed brick, the living room's second fireplace is handled in a less formal manner.

These darkly stained trusses are reminiscent of rice barn timbers (and airplane struts).

The Canadian goose was taken by the owner's wife. "Goose Marsh" was the unusual

1730 designation for one of the plantation's parcels (left).

French doors open this dining room onto a grand marsh vista. The table centerpiece

is a pair of canvas backs shot by the owner (above).

previous spread / Sunset (page 246, top). **A** *rice field surface turns to silver in the late afternoon* (page 246, bottom). **T**he folk-art weather vane was original to the building. *That's a peaceful Maine chief transported here to the land of marauding Yemassee* (page 247).

this spread / A *recent addition, this guest house is reminiscent of the Caribbean style once used by antebellum planters* (above, left). **L**ike the 1929 lodge, *this newly constructed guest house also suggests the material and craft of an earlier generation. Heart-pine paneling is on the walls, and the bath has a beaded Victorian finish* (above, right). **T**he surrounding porch of the guest house offers more than shade—it comes with a view (right).

CHEROKEE PLANTATION / House circa 1930s, Combahee River

Combahee, Yemassee, Edisto, Pocotaligo, and Ashepoo—this neighborhood to the south of Charleston is commemorated with Indian names, hundreds of them. But this plantation takes its name from the Cherokee roses growing on the site in 1930. And then more recently, the plantation name was given to the popular automobile the Jeep Cherokee.

Cherokee and her rice fields, with the river beyond (below) */ Photograph by Alice Turner Michalak.*
In 1930, this palatial brick home replaced the humbler Board House Plantation (bottom).
This massive front door is actually two medieval doors joined together (right).

The original, and far more prosaic, Board House Plantation was held for three centuries by the Blake family. Of these, Daniel Blake had the longest and most profitable tenure. He lived at Board House from 1803 to 1873, directing his planting and summering with his wife in the North Carolina mountains.

In 1860, he reported producing well over 1 million pounds of rice, 5,000 bushels of corn, 400 pounds of wool, 50 bushels of peas and beans, 100 bushels of sweet potatoes, and 3 tons of hay. This he grew, tended, and harvested on 11,000 acres of land with the labor of 559 slaves.

Blake was complimented by a Methodist missionary for instructing his slaves and building them a church where he would sit with them through the service. "A most humane master," wrote missionary Luther Banks.

Board House met the fate of most others, with dwelling house, out buildings, stables, and pounding, threshing and sawmills all burned.

By the time of his death in 1873, Blake had disposed of most of his other properties, and the Coe family of New York purchased Board House and made it their winter home.

previous spread / The rice displayed on this richly carved Scandinavian mantel
was grown on Cherokee last year. The rapidly moving horse and jockey is a
St. James Club trophy (page 252, top). **T**he grand broken pediment surround
is of Scandinavian pine. The banquet table seats 16. The mural on the far wall
celebrates a deer hunt and was painted by Carele Vernet in Paris in 1815 (page 252, bottom).
A Louis XVI gilded cabinet is the room's centerpiece, but seen through the window
is an equally momentous view of the river and rice fields (page 253).
this spread / The Italian oak paneling for the library was taken from an English castle.
The framed mirror was found with glass turned to the wall in the attic above. The lady over the
sofa dates from the Coe ownership, as well. The partner's desk is leather topped (left).
The garland of this hand-carved overmantel features quail and Cherokee roses. And the distinctive
powder horns are actually a part of the imported marble mantel (above).

previous spread / A *gate at sunset* (pages 256–257).

this spread / A *1926 Thames River slipper boat. And the 1926 "Herring Gull"—a "J" class tender used by the Vanderbilt family to watch the Newport yacht races* (left). **B***uilt in the '30s, the first boat house burned. This handsome red-cedar construction is a recent replacement* (above).

following spread / A *fitting symbol for this extraordinary estate. Frederick Law Olmsted's firm (of Central Park fame) designed these grounds* (page 260, top). **T***he golf clubhouse is as unobtrusive as the course itself, which was designed by award-winning Donald Steel* (page 260, bottom). **N***ot so long ago, rare Peruvian Paso horses were bred here* (page 261, top). **A** *length of evening shadows* (page 261, bottom).

Being an avid horticulturist, W.R. Coe renamed the plantation for the Cherokee rose. He hired the famous Olmsted landscaping firm to draw up a design plan for the entire acreage. Then Coe gathered together the rooms from several European houses and castles and had architects design his own handsome mansion around those.

In the early '50s, Cherokee passed into the hands of R.L. Huffines, an industrialist and board member of ABC television. Huffines then sold the property to Bob Evans, the president of American Motors. Evans took the plantation name for his redesigned Jeep.

After two more owners, Peter de Savary arrived at Cherokee with a dream of preserving Cherokee's unique qualities and sharing its genteel plantation lifestyle. Not a developer in the usual sense of the word, the new owner approached this project from the direction of hospitality— creating a beautiful, quiet, rural retreat for a particular and very limited clientele.

Tulip magnolias showcase the springtime riches at Cherokee (below). **O**ne of many amenities (right).

CLARENDON PLANTATION / HOUSE CIRCA 1934, WHALE BRANCH

Clarendon Plantation appears first in the records as a 350-acre parcel being sold by John and Elizabeth Fyler. But documentation is scarce, and the 350-acre parcel appears in an 1865 deed transfer. Fyler, a Beaufort merchant, had sided with the Union and then taken his wife to New York City so he did not, like most of his neighbors, lose the plantation for unpaid taxes.

---

*O*ak and azalea in complement of architect Willis Irvin *(below).* M*arble and wrought iron greet the guest* (bottom). T*he house features a double-tiered portico in the Federalist manner—one so popular in nearby Beaufort, it's called the Beaufort style* (right).

Those who came next attempted to grow cotton. In the 1880s, the owners switched to truck farming. But by the Depression-saddled 1930s, even this failed.

Modern Clarendon Plantation was formed when brothers Warren and Henry Corning purchased six plantations, which they named Clarendon.

In 1934, Henry had well-known Augusta architect Willis Irvin design a home. The house, built in the great country house manner, was to serve not only as a year-round residence but as a hunting lodge. In addition, Umberto Innocenti, a popular and skilled landscape designer, designed a riverside lawn and formal garden and an entry court on the house's land side.

The Corning brothers sold their holdings in 1944, and a dairy and beef cattle operation was carried on there through the '60s.

In the most recent years, the land has been used relatively lightly, a policy that has been consciously extended. The owners are proud that they have been able to retain their plantation relatively unchanged in the midst of rapid residential development on every side.

As to the very earliest days, an interesting

The broad sweep of rail around the stair's newel is a hallmark of
popular architect Willis Irvin, as is the broad dentil above. A pair of late 19th-century Ferahan
runners lead the visitor inward (left). **A** view of the rear of the house with azaleas (top).
Famed landscape designer Umberto Innocenti created this formal garden (above).

study of Clarendon's many cemeteries was done by preservation consultant Sarah Fick. Fick's examinations of the burial records and existing tombstones suggest a surprisingly broad range of occupants, as well as an alarming mortality rate. The first-known burial was in 1756—a young son of Beaufort contractor Henry Tailbird. And in early 1782, Capt. James Doharty was interred after being killed by British loyalists. Then several more of the Tailbird family joined these two. John Rhodes and his wife, Mary, are buried at another site. The most prominent is Paul Hamilton, wealthy rice planter, governor of the state, and secretary of the U.S. Navy. But perhaps Mary Rhodes' husband Nathaniel did as well. Born on this property, he had received a medical degree from the University of Edinburgh in Scotland. He fathered eight children and then in 1817 died in Beaufort while tending the victims of a yellow fever epidemic. One in six of the white population died, including Nathaniel and Paul Hamilton, Jr. And in later years, Nathaniel's son John, also a medical graduate, would die of fever at the age of 21. ❧

This impressive vaulted ceiling was borrowed from a Combahee River rice barn (below). This comfortable room is highlighted by a "coffee table" built from a Victorian lacquer and gold chinoiserie-decorated papier-mâché tray on a faux bamboo stand. The end table is derived from library steps from the Regency period (right).

GREGORIE NECK PLANTATION / HOUSE CIRCA 1930S, COOSAWHATCHIE AND TULIFINNEY RIVERS

The records are sketchy, but this neck of land between the Coosawhatchie and Tulifinney rivers and south of "the public road" was purchased by Alexander Gregorie from a Mrs. DeVeaux in 1798. The name shifted appropriately then from DeVeaux's Neck to the current name.

Gregorie divided the land between three of his children. The houses they built were all burned at the end of the Civil War.

After several changes in ownership, Bayard Dominick bought the property and, with the aid of well-known architect Willis Irvin, constructed the present house. Of brick and with a maximum exposure to the riverside view, the stately home remains little changed today.

The present owners use the plantation not only for hunting but as a family weekend retreat. ✿

A *misty morning light.*
Simplicity with distinction (below).
Cool breezes from the river
sweep through
the hallway out to the
shaded front lawn (right).

previous spread / **M**ementos of a hunting life. The pouncing owl was done by
Savannah carver and naturalist Floyd Robbins. The shotgun in miniature is of sterling silver
and was presented to the owner by Dr. Ugo G. Beretta of Beretta Arms (page 272, top).
In the gun room, the wall treatment of faux maple burl was painted by Kipling Collins
of Savannah. The turkey and deer were taken at Gregorie Neck (page 272, bottom).
Above the mantle in the living room is a map of the plantation
painted by artist Floyd Robbins (page 273).
this spread / **T**his English mahogany pedestal table was among the furnishings purchased
with the house. The oil paintings are by well-known American Impressionist Harry L. Hoffman
(1871–1964) and depict plantation scenes in the Savannah area (left).
This modern-day "butler's" pantry is a perfect transition from kitchen to dining room.
A well-worn kitchen table got a new butcher-block top. On the far wall is a late 18th-century
"mule" chest now converted to a dresser. The china reflected in the mirror is hunting dogs
and Spode "Woodlands" (above, left). **A**n avid hunter, the owner built this
recent addition to house trophies from around the world (above, right).

previous spread / **A** *broad extension, roofed and screened, helps to bring the outside in or vice versa* (page 276, top). **O**n point. "The little bobwhite," as one fan explains, "came to represent the nobility of the animal world; and those who hunted them, the aristocrats of hunters (page 276, bottom)." **T**he original gift to the owners was a single pair of mallards—now swollen to a flock and all content to stay and paddle (page 277, top). **A**nother stately home by architect Willis Irvin. River views and river breezes got full consideration (page 277, bottom). **this spread** / "**T**he dogs love the truck," says the owner's wife. "They're a big fraternity of dogs who love to ride (above)." **M**orning haze on the Tulifinney River (right).*

CHELSEA PLANTATION / House Circa 1937, May River

John Heyward's father died when he was 2, and he was raised by his uncle Thomas Heyward, Jr., a signer of the Declaration of Independence and successful planter. When John married in 1827, Thomas gave John and his new wife 3,000 acres with a substantial house built on the property.

Upon returning from their two-year honeymoon tour of Europe, the new owners chose the name Chelsea—for Chelsea, England, held for them the most romantic memories of their journey.

this spread / Antebellum avenue and Depression-era gate (below). Charleston architects Simons and Lapham created this graceful elliptical and most Charleston of arches (right).

following spread / Camellia-laden fabric and azalea-laden vases (page 282, top). Chippendale mahogany side chairs. An English painting of pointers among the sand dunes is against a wallpaper of hunting print (page 282, bottom). A broad range of fabric and plenty of sunlight brighten this spot. Above the mantel are a pair of English setters and a dog of mystery. The painting is English (page 283).

This couple had one daughter, who in turn married and had one daughter, who lost the property after the Civil War. The house was spared by Union troops. Northern hunters bought the place for quail hunting and added to their holding until Chelsea had swollen to 18,000 acres. In the early 1930s, the house burned. Marshall Fields III bought the property in 1937, and with a design by Simons and Lapham of Charleston, built the present house.

Today's owners maintain and use Chelsea for quail hunting. ≈

previous spread / **P**ines and azaleas (page 284, top). **R**eflections from a tree-shaped
pool (page 284, bottom). **A**ncient oaks suggest to the house a much greater age (page 285, top).
Spring tide on "The River the French called the May" (page 285, bottom).
this spread / **M**orning light (top). **A**rchitects Simons and Lapham were masters of the antebellum
plantation style. The site was established long ago for a honeymooning couple (above).
Simple details form a rear entry (right).

AULDBRASS PLANTATION / House circa 1938, Combahee River

In belated recognition of his half-century of revolutionary "organic" designs, architect Frank Lloyd Wright appeared on a 1938 cover of *Time* magazine. In that same year, he began work on an Auldbrass Plantation house for New York economist and businessman C. Leigh Stevens.

Stevens had acquired 4,000 acres of rice fields and forest named in honor of a slave "Old Brass." A slight shift in name had taken place. And now Wright was asked to design for a modern self-sufficient farm, a commission that would involve not only a house but an entire complex of farm-related buildings.

By this date, the aging Wright sometimes allowed his young associates to make substantial contributions, but Auldbrass got his personal attention. More than 500 pieces of correspondence exist regarding the construction. At times, this hands-on approach led to frustrating clashes over just what constituted beauty—and convenience.

The main problem, however, was beyond the control of either the owner or architect: With the beginning of World War II, materials and labor were increasingly difficult to acquire. Major concessions had to be made, especially in the use of copper. Much of the complex according to the original drawings and specifications was never constructed.

The Auldbrass logo—an update from the antebellum "Old Brass," which was named in honor of a slave (below). *This pool was recently installed. It was part of Wright's 1938 plan for the property* (right).

Nevertheless, Auldbrass, a striking assemblage of long, low components—of concrete, cypress, glass, and copper—was completed to a degree and enjoyed for several decades.

In 1979, the family sold the property and a process of abandonment began. Several buildings had already burned, and the roofs on those remaining started to leak. Eventually, a private hunting club camped in the rapidly deteriorating main house.

Fortunately for this treasure of a house, relief came in 1986.

A new owner, a well-known movie producer

*previous spread / **T**his entry arch prepares the visitor for the visual treats to come* (page 290,

top & bottom). **W***right's Midwestern predilection for the low horizontal served equally well*

in a land of azaleas and live oaks (page 291, top). **W***ith broad bases, tapering walls, and distinctive*

heavily patinated copper roofs, the buildings, themselves, are a grove of oaks (page 291, bottom).

*this spread / **C**ypress, glass, and welcoming light* (left).

O*f particular interest is the stylized copper moss dripping from the corner* (above).

*following spread / **A** moose and antique tiger offer a playful commentary on hunting-lodge*

aesthetics. For Wright, the hearth was the center of the home. And from the brick above this one,

a hidden cantilevered steel beam literally raises the roof. Note beneath the tiger's chin.

That 5-foot hexagon grid is repeated in the floor throughout and everything else

from walls to furniture takes it cue accordingly (pages 294–295).

previous spread / **T**wo Wright-designed chairs wait beneath a movie projector. Surprisingly, two of the more traditional plantation homes in this neighborhood contain complete movie theaters (page 296, top). **T**his office space occupies what was once a gun room and houses part of the owner's extensive toy collection (page 296, bottom). **A**bove are four famous clowns and two train models, one propeller driven, that engineers constructed in the 1920s. Toys, movie videos, a Tiffany lamp, and a glimpse of the lake round out the decorations (page 297). this spread / **G**lass on both sides, and at the center, table(s) and chairs designed by Wright. A blaze of colorful dinnerware seems to steal the show (left). **B**oathouse (above). following spread / **Y**oung cypress well on their way to a majestic maturity (page 300). **A** guest house at twilight (page 301).

with a fondness for all things Wright, took the architect's original drawings and specifications and began restoring the plantation.

The existing buildings were repaired and finished to the intended luster, and those lost to fire and rot were replaced. Now those planned, but unbuilt, structures of the compound are being constructed.

A fair amount has been written about the house's beauty, for critics and scholars are finding their way to the remote site.

this spread / T*his beer girl was a design that Wright used in Chicago's Midway Garden. The building is gone but not the spirit* (below). **T***his light-filled clerestory circles the house and creates a ceiling that almost floats* (right). **following spread / D***esigned to be a private resort and working farm, Auldbrass is gradually being restored and completed* (pages 304–305).

What they say has relevance for the plantation houses being built today and those of the future. Above all, Wright believed that architecture should be connected to nature, that the windows and doors open dramatically to the outside world, especially along the sun-filled exposures, that the silly ornamentation of the Victorian era be abandoned—as well as the conventional demands of Georgian's symmetrical organization.

Kitchens were allowed to flow into the more formal areas of the house. And as at Auldbrass, each space was meant to grow from the one preceding and fold into the one that follows.

Such construction called for increasing use of steel, concrete, and glass—materials embellished here with a richly colored cypress and patinated copper. The patina was acquired with a guarded Wright formula and was meant to complement the oaks.

Above all, the concern is with the sense of shelter, often dramatic, but at the same time comforting and secure.

While the traditional forms of plantation architecture may still hold sway today, Wright's notions of harmony, space, and light are being adapted to Lowcountry uses. 🙵

R obert Marvin is considered by many to be the South's most influential landscape architect. A half-century ago, he worked with Charles Fraser to set the development standards for Hilton Head Island. Marvin's nature-based aesthetic would influence all that followed. "When my wife, Anna Lou, and I came back here after school (in the late '40s), we had an idea that someday the

What Charles Fraser says, 'Whales were planned by God to live in the ocean,' is true. Look at it this way. Man and woman were built by God to live in a garden. Therefore cities should be gardens ... even corporate high rises should be gardens ... which is going to require a new kind of architect, a new form of architecture.

My grandfather planted a thousand acres of rice. And at the end of the day, he needed a cave to go

South would get rich enough to mess itself up. We in the South had a wonderful landscape and there were people with the right values and work ethic, the right morals to preserve it—if given the chance. And you know the South in my lifetime has had that wealth come, and if we hadn't planned, our corner of it would have been ruined like so much of the rest of this country.

I'm happy to have stayed right here, to have not expanded with the company beyond the South. We could have gone to England. We even turned down an opportunity in Dallas. Ego would have been the only reason for leaving. I didn't care about making money. The land was what mattered—the doing good.

To design badly is a sin. (Oh, almost a sin.) You look to the site, design to the land. A landscape architect should be on every team and have powers equal to the architects. But back then this wouldn't have happened in the South—or outside of it very often. We weren't producing any Thomas Jeffersons in this country. Life had gotten way too complicated. People suffered from tunnel vision. When I got out of school in the late '40s, architecture and landscaping students weren't being prepared with a vision of a total world. And clients weren't prepared to accept such a vision either.

This ability to see and do the whole thing. ... Before World War II, anything built, anything from a privy on up, was usually built better than it had to be. Projects were approached as important. After the war, that was changing. People were in a bigger hurry and they were losing contact.

into. His house, most plantation houses, were just boxes. Some had big windows, some were quite impressive, even beautiful in their way—but still they were boxes. The technology of his age had cut out all the rhythms of nature. We, today, need to be conscious of the sun. We need to knock down those walls and let the light in. Our emotional well-being and that of our kids and grandkids should be tied to nature. I can't say my grandfather needed this. He was outside every day all day long. He was trying to raise crops. He was in an adversarial relationship with nature. But that's behind us now.

I grew up on a plantation. On Bonny Donne over by Ritter. They had a garden designed there by the famous landscaper Umberto. This was a big place, a hunt club. When I went to catch the school bus, I was driven 7 miles. I probably learned more from the black man driving me that 7 miles than from anyone else in my life. I learned more about being a human being. My father was the manager. I had six brothers and sisters. We all went to college. I got a degree in horticulture from Clemson then went to the University of Georgia for an art degree, but didn't graduate. I didn't make good grades, but still I came away with a vision of what landscaping entailed. So once out of school ... beginning around 1955, I began to educate myself. I knew little about art. But I knew art and not science should be the basis of my business. For three solid years, I read and talked to people, and I began to understand the principals of design, to understand color and form, texture.

I saw how simple it was to get a house to blend in—to put a couple of trees in front. I had a talent. Now the Bible says you can't serve two masters. Consider an art object. It has a dominant element. I couldn't make money and be the best landscaper that was possible. My wife wrote us a motto. 'Robert Marvin and Associates' goal would be to create an environment in which each individual could develop into a full human being as God intended him or her to be.'

The word environment ... I don't think I'd ever heard it spoken in South Carolina. But in 1962, my wife and I drove in our Volkswagen out to a conference on the environment in Aspen, Colorado. We didn't have the money to fly. Two thousand people from all around the world showed up. Jonas Salk and others led. Artists, businessmen, scientists—every discipline was represented. Ten days spent there in this huge orange tent on the side of a mountain, the orange tent in a field of wildflowers and standing against the blue background of the sky. They spoke of how Americans could only react to their physical environment—to the material world. But their emotional environment was what affected their happiness, and they didn't even know how to talk about it. They lacked the vocabulary. Before that I'm not even sure I'd heard the word ecology spoken outside of a classroom. We came home invigorated.

And since then we've been studying, traveling, working ... coming out of the 'tunnel.' I understood that an individual was the product of two forces—the God-given characteristics that made them special and the environment that surrounded him. And that environment should allow him to grow in every positive way possible. This is true for a house. It's true for the center of a city—and in the past decade much of my work has been in the cities.

But how does a designer begin? If it's a residence, he goes to visit the family. You visit the kids. Three kids equals three personalities. Plus the parents. Two more personalities. They become a part of you. To be precise, you bond. Then you get out on the land. You have five different maps to reflect vegetation, slopes, soils, and physical characteristics—is there noise? where are the neighbors? how does the sun move? Mark all the trees. You build with the land. You don't cut the trees. Develop a house suited

to the site, not vice versa. The land comes first.

The plantations today? I've spoken already of the need to let light into the house. To abandon that adversarial relationship with nature that was my farming grandfather's lot. All day long he fought an unpredictable nature. We have it still with us today, that insecurity that requires man to dominate nature and retreat at night into a box. But not the justification. The gardens and immediate grounds can be considered in the same light. The live oaks, the camellias and azaleas, they harmonize. The oak avenues are classical—an ordering. You know, it's hard to talk about the emotional needs that a garden involves. In Europe with a king and queen, you had a hierarchical world. Everything was formal, walled in. Trees got moved. Boxwoods got clipped. Then William Kent jumped the fence and discovered that all nature was a garden. Arts went informal then. And for a democracy like America that should have been the answer. But in those early days, we were in a wilderness ... and emotionally there was usually a need for control in the garden, for trimming and geometrical security. Again, I think we have passed that.

In the Lowcountry, the wealth has always been in the plantations. In the beginning, rice was king. And indigo and cotton came along. But the Civil War ended that. Sherman burned every house around here. By the time of the Depression, all the plantations had gone broke. The people who owned them might stay on as superintendents—my father was one. But at least the new owners cared about the land and they had the money to maintain it. Then in more recent years, the plantations were being sold once more. They were too expensive to maintain or the next generation had lost interest. Timber companies were buying them. ... But some property owners continued to care. Ones like Gaylord Donnelley, a wonderful man who fell in love with the land. He said, I can't let this be destroyed. He invited the governor to go duck hunting. He involved people, and the result is the ACE Basin project—easements that preserve. Just remember. The land comes first." 🔸

MACKEY POINT PLANTATION / House circa 1997, Pocotaligo River

This point of land between the Pocotaligo and Tulifinney rivers appears on the earliest of maps as "Indian Fields." And no doubt, this house site was previously an aboriginal settlement. A Mr. Hudson was shown as the owner in 1732, but the Union's burning of courthouse records has erased most else.

An 1806 newspaper advertisement for the property begins on a confident note:

_____

This new house is actually built on a recently constructed promontory (below). **A** *more contemporary avenue of oaks* (bottom). *The brick columns just beyond the glass serve to psychologically separate in from out* (right).

"The celebrity of this place renders a particular description almost unnecessary. Suffice it to say that the Lands are almost entirely of the first quality for the culture of cotton, and its local advantages by being the lower part of the neck, must be obvious ..."

The writer goes on the say that all but 270 recently cleared acres of the thousand plus is wooded, that there are several boat landings, and a good two-story house and large barn have been recently built.

In 1862, Robert E. Lee, then in charge of coastal defenses, was reported to have stayed with the Mackay family.

And when Sherman, the other visiting general, came through, he burned down the house and outbuildings.

Eventually, George D. Widener bought the plantation as a shooting preserve and constructed a rambling Colonial lodge on the site.

The present owners found this long-neglected building to be beyond repair and removed it so they could start over fresh.

The firm of Thomas and Denzinger was hired to design and supervise a main house, guest house, and river house—and all of these

previous spread / **A** *fireplace and especially dark paneling create a sense of shady retreat in this screened porch great room* (page 310, top). **P***rerequisites to relaxation* (page 310, bottom). **T***hese heart-pine lockers were inspired by those of a nearby and much earlier hunt club. Each hunter has his own, and the felt-covered worktable in the center is shared by all* (page 311).

this spread / **A***n old-time fish camp was the inspiration for this river house. The heart-pine paneling and beams inside the river house received a brightening coat of milk white wash* (left). **T***his lengthy boardwalk protects the vegetation along the river's "critical line." Regulations prevent the injury of these productive wetland areas* (above).

following spread / **C***lose by the river and relying a bit on nature's landscaping, this modern-day fish camp seems to turn the clock back* (page 314, top). **A** *view of the new house and an already established garden* (page 314, middle). **T***wilight and oaks. The main house presents a traditional face to arriving visitors* (page 314, bottom). **T***he gold of dusk finds its way inside* (page 315).

buildings were required to make better use of the property's incredible riverside views. Under the initial guidance of Jim Thomas, architect Joel Newman set out to design accordingly. He drew loosely on the Adirondack hunting-lodge style, which had been used on neighboring hunt clubs earlier in this century—a functional, steep roofed form of shingle architecture.

To refurbish the "old" house site, landscaper Frances Parker simply removed all from the lawn except the live oaks. And the original garden is now defined by a brick wall enclosure that connects the main house to these guest quarters (below).
Gold sprinkles the ground as sun sets over the marshes (bottom).
This auxiliary "cottage" is set for Santa's visit (right).

But the antebellum Lowcountry still offered inspiration for the architects, most obviously in the main house with its round brick "Sheldon church" columns.

Following in the tradition of the noted landscape architect Robert Marvin, Newman's designs stressed a modern relationship to the "outside."

While the 19th-century planter might, with some justification, see nature as his enemy, we have the comfort technologies and the leisure to do otherwise. More windows were needed, especially along the riverside, and the sun and breeze were invited in.

And because the property was actually owned by two business partners and was to be used by them and their families on a shared basis, an unusual H-shaped layout was prescribed for both the living areas of the first floor and the bedrooms of the second.

The cross bar of the H served on the first floor as a linking service area and on the second floor as a common study that could be used by both families.

Now being enjoyed by the owners, the resulting structures are comfortably containing and yet firmly connected to an immense outdoors. ❧

PLUM HILL PLANTATION / House circa 1996, Combahee River

Being the youngest of pioneer Daniel Heyward's three sons, Nathaniel inherited little rice field wealth, but in the course of his life, he worked diligently and accumulated an immense 25,000 acres. Today's Plum Hill was carved from that whole two centuries later—with a more evenhanded distribution in mind. The intent of the present owners was to create not only a retirement home for themselves, but

The columns are reminiscent of nearby Sheldon Church ruins. As architect Virginia Lane explains, the "see through entrance ... orients you to the whole living experience." And so does the faithful guardian (below).

This single multipurpose great room serves as a center for the house and in the future it will act as a congregation point when the next generation of the family builds houses on the surrounding property. Note how the light enters from above through both dormers and a banding of glass or clerestory (right).

a vacation spot for their children. The house that architect Virginia Lane designed for them is essentially sleeping spaces centered around one large room, a high-ceilinged bright room that can accommodate from one to many, a place to cook, entertain, or simply relax. The owners were specific about their need to connect to the surrounding land. For materials, they were particularly helpful, locating the heart-pine beams, cypress, and old handmade bricks that suggested an even further connection to the land and its history. The result is a very livable, contemporary / traditional plantation home.

Centered on the hearth is the tried-and-true goal of both traditional and modern architecture, and because this house is a combination of both, it's not surprising to find this handsome arrangement. A broad grouping of windows provides an unimpeded view of the river and marshlands (left). **A** porch with a view. Sunset on the Combahee River (top).

Seen from the adjoining fields. The "car port" roof is supported by the distinctive columns, which, of course, lend the pavilion to other uses, both informal and otherwise. Last spring it served quite successfully as a wedding chapel. Fishing is great, too (above).

The distinctive "Sheldon Church columns" suggest a continuity with the plantation past (top, left).
Two doors and a wide serving window prevent the isolation of the kitchen. And the warm glow of
heart-pine and cypress panels connects the rooms, as well (top, right). **H**ere a wide, sunny hall
does double duty as a library. Note the strutting turkey gobler. And note how this
interior vista is extended even farther by ending with a bedroom window.
Such devices were used throughout to not only connect inside to out,
but to create an even more spacious feel (right).

Virgina Lane, a Charleston-based architect, has close ties to the Lowcountry—and especially to both its land use and building traditions. ¶ "The owners first approached me saying they wanted to build a big room in the country. They had seven children and many grandchildren, and they wanted a room where they could entertain and cook and live and retire that was flexible enough to the outside. And the 'outside' to me became many things—and on a lot of levels I started working with that. How would this room relate to the beautiful site on the river? To the incredible mirroring black water of the Combahee River? And the farm on the other side? The husband loves to farm and does it quite seriously. He's a surgeon. And the spaces he creates by building on the landscape—well, the question was how do you frame the edges of the exterior spaces.

I'm trained as a modernist and truly love modern architecture. But context and history of this building forms in South Carolina in particular are exquisite and the simple pure forms of the 18th century are hard to beat. They seem to be so well rooted. And the materials, too. So a lot of the building reflects that. The round brick columns clearly standout. We started with this large room and added wrap-around porches in the manner of Otranto Plantation. We looked at that and after drawing it up several times, realized the amount of waste these porches can create. The main room does still expand out into a big porch on the river side and a porch on the front side—and into a den adjacent. But for a form we moved from the wrap-around and looked at Old Sheldon church because that's the closest ruin in that area. Then we looked to Medway, the oldest brick house in the state, and the three of them do have in common the brick—and the brick was made on the plantation.

Material was the second part. The materials, the owners were wonderful about that. They located these incredible heart pine beams, the natural cypress from a local mill, and they found the bricks. And, you know, it has more meaning that way. A happy marriage of all involved. The concept of the see-through entrance I enjoy. You go through a procession of entry spaces and finally you get to that front door. You can see completely through the house to that beautiful river view on the other side. It orients you for the whole living experience. That was the organizing factor, there—that axis. The breaking down of the other spaces—their bedroom, two guest bedrooms, breaking that scale down so that it wasn't massive—wasn't a 'mega-home' was more on the scale of a small home. That's the truth in so many additions, so many wings. You need to break down the masses—the central mass being the living room.

It's all on one floor. They wanted that because it was a retirement home. The dormers and the clerestory above make it look like a two-story space, almost like the main plantation house and then the smaller ones are like out buildings.

With their seven children, each has the potential for building a house on the property, so this first house has some of the responsibility — being the patriarch of the farm. We needed to succeed without creating a mega-home. The easement on the property requires that each house be under 3,000 square feet. And we are that. It only looks bigger.

My husband has worked for many years on the ACE basin effort and that made the house all the more special. I was working with Thomas and Denzinger Architects when I did this project. Jim Thomas took me under his wing. And the contractor was terrific. Jim Thomas told me I'd have to pitch a tent on the site to get this built—between the round columns and the gable ends—and those were important—the Medway influence. But the masons and the contractors were extraordinary. I'd love to work with them again." ⁂

A perfect melding of traditional and contemporary design.
Plum Hill house with long shadows (right).

DELTA PLANTATION / House circa 1900, Savannah River

The house, built around 1900, is a copy of Tidewater Virginia's Brandon Plantation designed by Thomas Jefferson. The classic facade, a two-story block connected to one-and-a-half-story wings, boasts long enfilades and rooms of original and exquisite wood paneling. The owners realized that this arrangement would lend itself perfectly to the kind of entertaining they wanted to have in association with hunting.

The current owners of Delta Plantation, both of whom are natives of Savannah, were living in England when they purchased this historic rice plantation. The furniture was bought in England and Scotland; fabrics are old; and the whole was intended to look as though it had always been there, showing gentle wear and loving use by successive generations.

Delta Plantation on the Savannah River began officially in 1830 when Langdon Cheves combined several existing tracts. The large and efficient rice plantation that resulted was considered the finest in the area and a model for others to come—a model of the "frontier" plantation.

While rice had been grown in the Lowcountry for more than a century, these Savannah river marshes had been largely ignored, in part because of their reputation for deadly fevers—a well-deserved reputation.

Balance and order greet the visitor (below).
To greet the visitor is this Irish hall table above which hangs an exotic bird case of the Victorian era. Locally collected shed snake skins add a bit of local color (right).

Though Cheves built a modest residence on the spot, this was essentially for the overseer. White owners seldom, if ever, spent the night on the river.

Even the white overseers preferred to sleep in Savannah and be paddled to work in the morning.

And while the slaves had some genetically instilled protection from malaria, cholera and other diseases took a grim toll. Indeed, Delta was remarked on as being one of the only plantation on the river where the slave population could sustain itself—where the successful births equalled deaths.

Much happier is the present situation. 🐌

The Duchess of Devonshire (a stylish 19th-century reproduction in a handsome,

limed period carved frame) looks down on an English large round table of sun-bleached mahogany.

The chairs are London made, each linen slipcover hand painted by local artist

Elsie Tallifaro Hill with the Delta logo—a wild boar and a latin quote that roughly translates:

"Happy is the man who knows the woodland gods." The side board is actually an

English architect's table from 1815. The chandelier is French.

Seen through the door is a Morning Room centered with an 18th-century

English country table (left). From the river (top).

And from the land (above).

previous spread / All the accoutrements of shooting are displayed in this boot room. On the table a Beretta over and under awaits a cleaning. In the far corner, a "planter's chair" with extensions for the removing of boots, and close at hand, a lamp made from the vertebrae of a local alligator. Worn leather, pecky cypress, and a superb collection of antique decoys create a comfortable transitional space (page 330, top). **T**his is a professional working kitchen (page 330, bottom). **A**bove the mantel is a fine carved stag's head. The tables on either side of the fireplace are campaign furniture made for the Boer War. On the center table is a collection of silver boxes, inkwells, and match strikers (page 331). **this spread / I**n the master bedroom, the bed is 19th-century painted Italian, and the daybed is a century older and French. The French barometer above the mantel is late 18th century. The dressing table is an English pianoforte (left). **T**he roman shade's Fortuny fabric and the antique toile curtains make a distinctive backdrop for Mark Catesby prints and a mahogany washstand (above). **following spread / O**nce used to transport wine, this ancient amphora becomes a fountain; a golf course lies beyond (pages 334–335).